A Combinatorial Approach to Matrix Theory and Its Applications

DISCRETE MATHEMATICS AND ITS APPLICATIONS

Series Editor
Kenneth H. Rosen, Ph.D.

Juergen Bierbrauer, Introduction to Coding Theory

Francine Blanchet-Sadri, Algorithmic Combinatorics on Partial Words

Richard A. Brualdi and Dragoš Cvetković, A Combinatorial Approach to Matrix Theory and Its Applications

Kun-Mao Chao and Bang Ye Wu, Spanning Trees and Optimization Problems

Charalambos A. Charalambides, Enumerative Combinatorics

Henri Cohen, Gerhard Frey, et al., Handbook of Elliptic and Hyperelliptic Curve Cryptography

Charles J. Colbourn and Jeffrey H. Dinitz, Handbook of Combinatorial Designs, Second Edition

Martin Erickson and Anthony Vazzana, Introduction to Number Theory

Steven Furino, Ying Miao, and Jianxing Yin, Frames and Resolvable Designs: Uses, Constructions, and Existence

Randy Goldberg and Lance Riek, A Practical Handbook of Speech Coders

Jacob E. Goodman and Joseph O'Rourke, Handbook of Discrete and Computational Geometry, Second Edition

Jonathan L. Gross, Combinatorial Methods with Computer Applications

Jonathan L. Gross and Jay Yellen, Graph Theory and Its Applications, Second Edition

Jonathan L. Gross and Jay Yellen, Handbook of Graph Theory

Darrel R. Hankerson, Greg A. Harris, and Peter D. Johnson, Introduction to Information Theory and Data Compression, Second Edition

Daryl D. Harms, Miroslav Kraetzl, Charles J. Colbourn, and John S. Devitt, Network Reliability: Experiments with a Symbolic Algebra Environment

Leslie Hogben, Handbook of Linear Algebra

Derek F. Holt with Bettina Eick and Eamonn A. O'Brien, Handbook of Computational Group Theory

David M. Jackson and Terry I. Visentin, An Atlas of Smaller Maps in Orientable and Nonorientable Surfaces

Richard E. Klima, Neil P. Sigmon, and Ernest L. Stitzinger, Applications of Abstract Algebra with Maple™ and MATLAB®, Second Edition

Patrick Knupp and Kambiz Salari, Verification of Computer Codes in Computational Science and Engineering

Continued Titles

William Kocay and Donald L. Kreher, Graphs, Algorithms, and Optimization

Donald L. Kreher and Douglas R. Stinson, Combinatorial Algorithms: Generation Enumeration and Search

Charles C. Lindner and Christopher A. Rodgers, Design Theory

Hang T. Lau, A Java Library of Graph Algorithms and Optimization

Alfred J. Menezes, Paul C. van Oorschot, and Scott A. Vanstone, Handbook of Applied Cryptography

Richard A. Mollin, Algebraic Number Theory

Richard A. Mollin, Codes: The Guide to Secrecy from Ancient to Modern Times

Richard A. Mollin, Fundamental Number Theory with Applications, Second Edition

Richard A. Mollin, An Introduction to Cryptography, Second Edition

Richard A. Mollin, Quadratics

Richard A. Mollin, RSA and Public-Key Cryptography

Carlos J. Moreno and Samuel S. Wagstaff, Jr., Sums of Squares of Integers

Dingyi Pei, Authentication Codes and Combinatorial Designs

Kenneth H. Rosen, Handbook of Discrete and Combinatorial Mathematics

Douglas R. Shier and K.T. Wallenius, Applied Mathematical Modeling: A Multidisciplinary Approach

Jörn Steuding, Diophantine Analysis

Douglas R. Stinson, Cryptography: Theory and Practice, Third Edition

Roberto Togneri and Christopher J. deSilva, Fundamentals of Information Theory and Coding Design

W. D. Wallis, Introduction to Combinatorial Designs, Second Edition

Lawrence C. Washington, Elliptic Curves: Number Theory and Cryptography, Second Edition

DISCRETE MATHEMATICS AND ITS APPLICATIONS
Series Editor KENNETH H. ROSEN

A COMBINATORIAL APPROACH TO MATRIX THEORY AND ITS APPLICATIONS

RICHARD A. BRUALDI
UNIVERSITY OF WISCONSIN — MADISON
U.S.A.

DRAGOŠ CVETKOVIĆ
UNIVERSITY OF BELGRADE
SERBIA

CRC Press
Taylor & Francis Group
Boca Raton London New York

CRC Press is an imprint of the
Taylor & Francis Group, an **informa** business
A CHAPMAN & HALL BOOK

Chapman & Hall/CRC
Taylor & Francis Group
6000 Broken Sound Parkway NW, Suite 300
Boca Raton, FL 33487-2742

Printed in the United States of America on acid-free paper
10 9 8 7 6 5 4 3 2 1

International Standard Book Number-13: 978-1-4200-8223-4 (Hardcover)

Library of Congress Cataloging-in-Publication Data

Brualdi, Richard A.
A combinatorial approach to matrix theory and its applications / Richard A. Brualdi, Dragos Cvetkovic.
p. cm. -- (Discrete mathematics and its applications ; 52)
Includes bibliographical references and index.
ISBN 978-1-4200-8223-4 (hardback : alk. paper)
1. Matrices. 2. Combinatorial analysis. I. Cvetkovic, Dragoš M. II. Title. III. Series.

QA188.B778 2008
512.9'434--dc22 2008014627

Visit the Taylor & Francis Web site at
http://www.taylorandfrancis.com

and the CRC Press Web site at
http://www.crcpress.com

Contents

Preface

Matrix theory is a fundamental area of mathematics with applications not only to many branches of mathematics but also to science and engineering. Its connections to many different branches of mathematics include: (i) algebraic structures such as groups, fields, and vector spaces; (ii) combinatorics, including graphs and other discrete structures; and (iii) analysis, including systems of linear differential equations and functions of a matrix argument.

Generally, elementary (and some advanced) books on matrices ignore or only touch on the combinatorial or graph-theoretical connections with matrices. This is unfortunate in that these connections can be used to shed light on the subject, and to clarify and deepen one's understanding. In fact, a matrix and a (weighted) graph can each be regarded as different models of the same mathematical concept.

Most researchers in matrix theory, and most users of its methods, are aware of the importance of graphs in linear algebra. This can be seen from the great number of papers in which graph-theoretic methods for solving problems in linear algebra are used. Also, electrical engineers apply these methods in practical work. But, in most instances, the graph is considered as an auxiliary, but nonetheless very useful, tool for solving important problems.

This book differs from most other books on matrices in that the combinatorial, primarily graph-theoretic, tools are put in the forefront of the development of the theory. Graphs are used to explain and illuminate basic matrix constructions, formulas, computations, ideas, and results. Such an approach fosters a better understanding of many ideas of matrix theory and, in some instances, contributes to easier descriptions of them. The approach taken in this book should be of interest to mathematicians, electrical engineers, and other specialists in sciences such as chemistry and physics.

Each of us has written a previous book that is related to the present book:

I. R. A. Brualdi, H. J. Ryser, *Combinatorial Matrix Theory*, Cambridge: Cambridge University Press, 1991; reprinted 1992.

II. D. Cvetković, *Combinatorial Matrix Theory, with Applications to Electrical Engineering, Chemistry and Physics*, (in Serbian), Beograd: Naučna knjiga, 1980; 2nd ed. 1987.

This joint book came about as a result of a proposal from the second-named author (D.C.) to the first-named author (R.A.B.) to join in reworking and translating (parts of) his book (II). While that book—mainly the theoretical parts of it—has been used as a guide in preparing this book, the material has been rewritten in a major way with some new organization and with substantial new material added throughout. The stress in this book is on the combinatorial aspects of the topics treated; other aspects of the theory (e.g., algebraic and analytic) are described as much as necessary for the book to be reasonably self-contained and to provide some coherence. Some material that is rarely found in books at this level, for example, Geršgorin's theorem and its extensions, Kronecker product of matrices, and sign-nonsingular matrices and evaluation of the permanent, is included in the book.

Thus our goal in writing this book is to increase one's understanding of and intuition for the fundamentals of matrix theory, and its application to science, with the aid of combinatorial/graph-theoretic tools. The book is not written as a first course in linear algebra. It could be used in a special course in matrix theory for students who know the basics of vector spaces. More likely, this book could be used as a supplementary book for courses in matrix theory (or linear algebra). It could also be used as a book for an undergraduate seminar or as a book for self-study.

We now briefly describe the chapters of the book. In the first chapter we review the basics and terminology of graph theory, elementary counting formulas, fields, and vector spaces. It is expected that someone reading this book has a previous acquaintence with vector spaces. In Chapter 2 the algebra of matrices is explained, and the König digraph is introduced and then used in understanding and carrying out basic matrix operations. The short Chapter 3 is concerned with matrix powers and their description in terms of another digraph associated with a matrix.

In Chapter 4 we introduce the Coates digraph of a matrix and use it to give a graph-theoretic definition of the determinant. The fundamental properties of determinants are established using the Coates digraph. These include the Binet–Cauchy formula and the Laplace development of the determinant along a row or column. The classical formula for the determinant is also derived. Chapter 5 is concerned with matrix inverses and a graph-theoretic interpretation is given. In Chapter 6 we develop the elementary theory of solutions of systems of linear equations, including Cramer's rule, and show how the Coates digraph can be used to solve a linear system. Some brief mention is made of sparse matrices.

In Chapter 7 we study the eigenvalues, eigenvectors, and characteristic polynomial of a matrix. We give a combinatorial argument for the classical Cayley–Hamilton theorem and a very combinatorial proof of the Jordan canonical form of a matrix. Chapter 8 is about nonnegative matrices and their special properties that highly depend on their digraphs. We discuss, but do not prove, the important properties of nonnegative matrices that are part of the Perron–Frobenius theory. We also describe some basic properties of graph spectra. There are three unrelated topics in Chapter 9, namely, Kronecker products of matrices, eigenvalue inclusion regions, and the permanent of a matrix and its connection with sign-nonsingular matrices. In Chapter 10 we describe some applications in electrical engineering, physics, and chemistry.

Our hope is that this book will be useful for both students, teachers, and users of matrix theory.

Richard A. Brualdi

Dragoš Cvetković

Dedication

To Les and Carol Brualdi for keeping the family together

Richard A. Brualdi

To my grandchildren Nebojša and Katarina Cvetković

Dragoš Cvetković

Dedication

Chapter 1

Introduction

In this introductory chapter, we discuss ideas and results from combinatorics (especially graph theory) and algebra (fields and vector spaces) that will be used later. Analytical tools, as well as the elements of polynomial theory, which are sometimes used in this book, are not specifically mentioned or defined, believing, as we do, that the reader will be familiar with them. In accordance with the goals of this book, vector spaces are described in a very limited way. The emphasis of this book is on matrix theory and computation, and not on linear algebra in general.

The first two sections are devoted to the basic concepts of graph theory. In Section 1.1 (undirected) graphs are introduced while Section 1.2 is concerned with digraphs (directed graphs). Section 1.3 gives a short overview of permutations and combinations of finite sets, including their enumeration. The last two sections contain algebraic topics. Section 1.4 summarizes basic facts on fields while Section 1.5 reviews the basic structure of vector spaces of n-tuples over a field.

Matrices, the main objects of study in this book, will be introduced in the next chapter. They act on vector spaces but, together with many algebraic properties, contain much combinatorial, in particular, graph-theoretical, structure. In this book we exploit these combinatorial properties of matrices to present and explain many of their basic features.

1.1 Graphs

The basic notions of graph theory are very intuitive, and as a result we shall dispense with some formality in our explanations. Most of what follows consists of definitions and elementary properties.

Definition 1.1.1 A *graph* G consists of a finite set V of elements called *vertices* and a set E of unordered pairs of vertices called *edges.* The *order* of the graph G is the number $|V|$ of its vertices. If $\alpha = \{x, y\}$ is an edge, then α *joins* vertices x and y, and x and y are *vertices of the edge* α. If $x = y$, then α is a *loop.* A *subgraph* of G is a graph H with vertex set $W \subseteq V$ whose edges are some, possibly all, of the edges of G joining vertices in W. The subgraph H is a *induced subgraph* of G provided each edge of G that joins vertices in W is also an edge of H. The subgraph H is a *spanning subgraph* of G provided $W = V$, that is, provided H contains all the vertices of G (but not necessarily all the edges). A *multigraph* differs from a graph in that there may be several edges joining the same two vertices. Thus the edges of a multigraph form a multiset of pairs of vertices. A *weighted graph* is a graph in which each edge has an assigned *weight* (generally, a real or complex number). If all the weights of a graph G are positive integers, then the weighted graph could be regarded as a multigraph G' with the weight of an edge $\{x, y\}$ in G regarded as the number of edges in G' joining the vertices x and y. □

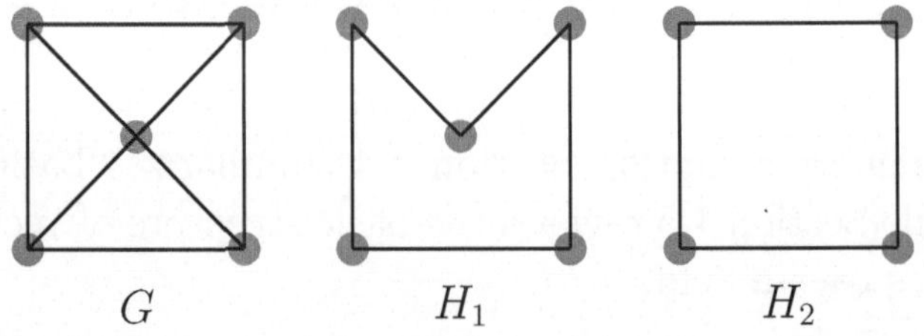

Figure 1.1

Graphs can be pictured geometrically by representing each vertex by a (geometric) point in the plane, and each edge by a (geometric) edge, that is, a straight line or curve joining corresponding geometric points. Care needs to be taken so that a geometric edge, except for its two endpoints, contains no other point representing a vertex of the graph. A graph G and two subgraphs H_1 and H_2 are drawn in Figure 1.1. The graph H_1 is a spanning subgraph of G; the graph H_2 is not a spanning subgraph but is an induced subgraph.

Definition 1.1.2 Let G be a graph. A *walk* in G, joining vertices u and v, is a sequence γ of vertices $u = x_0, x_1, \ldots, x_{k-1}, x_k = y$ such that $\{x_i, x_{i+1}\}$ is an edge for each $i = 0, 1, \ldots, k-1$. The *edges of the walk* γ are these k edges, and the *length* of γ is k. If $u = v$, then γ is a *closed walk.* If the vertices $x_0, x_1, \ldots, x_{k-1}, x_k$ are distinct, then γ is a *path* joining u and v. If $u = v$ and the vertices $x_0, x_1, \ldots, x_{k-1}, x_k$ are otherwise distinct, then γ is a *cycle.* The graph G is *connected* provided that for each pair of distinct vertices u and v there is a walk joining u and v. A graph that is not connected is called *disconnected.* □

It is to be noted that if there is a walk γ joining vertices u and v, then there is a path joining u and v. Such a path can be obtained from γ by eliminating cycles as they are formed in traversing γ. A path has one fewer edge than it has vertices. The number of vertices of a cycle equals the number of its edges. We sometimes regard a path (respectively, cycle) as a graph whose vertices are the vertices on the path (respectively, cycle) and whose edges are the edges of the path (respectively, cycle). A path with n vertices is denoted by P_n, and a cycle with n vertices is denoted by C_n.

Definition 1.1.3 Let G be a graph with vertex set V. Define $u \equiv v$ provided there is a walk joining u and v in G. Then it is easy to verify that this is an equivalence relation and thus V is partitioned into equivalence classes $V_1, V_2, \ldots, V_l$ whereby two vertices are joined by a walk in G if and only if they are in the same equivalence class. The subgraphs of G induced on the sets of vertices $V_1, V_2, \ldots, V_l$ are the *connected components* of G. The graph G is connected if and only if it has exactly one connected component.

A *tree* is a connected graph with no cycles. A *spanning tree* of G is a spanning subgraph of G that is a tree. Only connected graphs have a spanning tree, and a spanning tree can be obtained by recursively removing an edge of a cycle until no cycles remain. The graph in Figure 1.2 is a tree with 5 vertices and 4 edges. A *forest* is a graph each of whose connected components is a tree. □

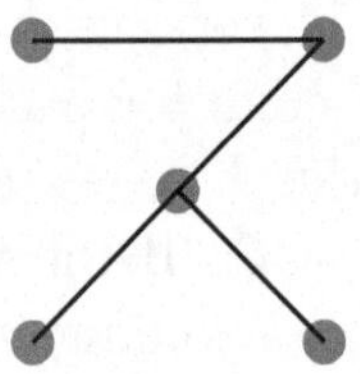

Figure 1.2

The next theorem contains some basic properties of trees.

Theorem 1.1.4 *Let G be a graph of order $n \geq 2$ without any loops. The following are equivalent:*

(i) *G is a tree.*

(ii) *For each pair of distinct vertices u and v there is a unique path joining u and v.*

(iii) *G is connected and has exactly $n - 1$ edges.*

(iv) *G is connected and removing an edge of G always results in a disconnected graph.*

□

An edge of a connected graph whose removal results in a disconnected graph is called a *bridge*. A bridge cannot be an edge of any cycle. Property (iv) above thus asserts that a graph is a tree if and only if it is connected and every edge is a bridge.

In Figure 1.3 we show all the structurally different trees of order k with $k \leq 5$.

Definition 1.1.5 In a graph G (or multigraph) the *degree* of a vertex u is the number of edges containing u where, in the case of a loop, there is a contribution of 2 to the degree. Let G be of order n, and let the degrees of its vertices be $d_1, d_2, \ldots, d_n$, where, without loss of generality, we may assume that $d_1 \geq d_2 \geq \cdots \geq d_n \geq 0$. Then $d_1, d_2, \ldots, d_n$ is the *degree sequence* of G. Since each edge contributes 1 to the degree of two vertices, or, in the case of loops, 2 to the degree of one vertex, we have

$$d_1 + d_2 + \cdots + d_n = 2e,$$

where e is the number of edges. A graph is *regular* provided each vertex has the same degree. If k is the common degree, then the graph is *regular of degree k*. A connected regular graph of degree 2 is a *circuit*. A *pendent vertex* of a graph is a vertex of degree 1. The unique edge containing a particular pendent vertex is a *pendent edge*. □

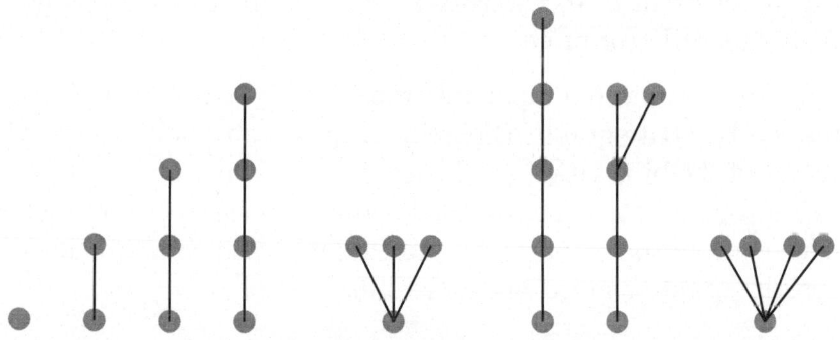

Figure 1.3

The *complete graph* K_n of order n is the graph in which each pair of distinct vertices forms an edge. Thus K_n is a regular graph of degree n and has exactly $n(n-1)/2$ edges. Since a tree of order n has $n-1$ edges, the sum of the degrees of its vertices equals $2(n-1)$. Thus a tree of order $n \geq 2$ has at least two pendent vertices, and indeed has exactly 2 pendent vertices if and only if it is a path. Removing a pendent vertex–pendent edge pair from a tree leaves a tree of order 1 less.

Definition 1.1.6 A *vertex-coloring* of a graph is an assignment of a color to each vertex so that vertices that are joined by an edge are colored differently. One way to color a graph is to assign a different color to each vertex. The *chromatic number* of a graph G is the smallest number $\chi(G)$ of colors needed to color its vertices. □

The chromatic number of the complete graph K_n equals n. The chromatic number of a circuit is 2 if it has even length and is 3 if it has odd length. The chromatic number of a tree of order $n \geq 2$ equals 2. This latter fact follows easily by induction on the order of a tree, by removing a pendent vertex–pendent edge pair.

Definition 1.1.7 A graph G is *bipartite* provided its chromatic number satisfies $\chi(G) \leq 2$. Only when G has no edges can the chromatic number of a bipartite graph be 1. Assume that G is a bipartite graph with vertex set V and at least one edge. Then V can be partitioned into two sets U and W such that each edge joins a vertex in U to a vertex in W. The pair U, W is called a *bipartition* of V (or of G). □

If G is a connected bipartite graph, its bipartition is unique. A tree is a bipartite graph. Bipartite graphs are usually drawn with one set of the bipartition on the left and the other on the right (or one on top and the other on the bottom); so edges go from left to right (or from top to bottom). An example of such a drawing of a bipartite graph is given in Figure 1.4.

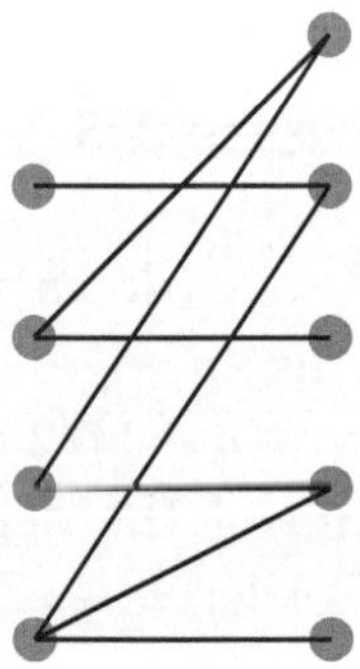

Figure 1.4

Let m and n be positive integers. The *complete bipartite graph* $K_{m,n}$ is the bipartite graph with vertex set $V = U \cup W$, where U contains m vertices and W contains n vertices and each pair $\{u, w\}$ where $u \in U$ and $w \in W$ is an edge of $K_{m,n}$. Thus $K_{m,n}$ has exactly mn edges.

Definition 1.1.8 Let G be a graph of order n. A *matching* M in G is a collection of edges no two of which have a vertex in common. If v is a vertex and there is an edge of M containing v, then v *meets the matching* M and the matching M *meets the vertex* v. A *perfect matching* of G, also called a 1-*factor*, is a matching that meets all vertices of G. The largest number of edges in a matching in G is the *matching number* $m(G)$. If G has at least one edge, then $1 \leq m(G) \leq \lfloor n/2 \rfloor$. A matching with k edges is called a k-*matching*.

A subset U of the vertices of G is a *vertex-cover* provided each edge of G has at least one of its vertices in U. The smallest number of vertices in a vertex-cover is the *cover number* $c(G)$ of G. If G has at least one edge that is not a loop, then $1 \leq c(G) \leq n - 1$.□

The complete bipartite graph $K_{m,n}$ has matching and covering number equal to $\min\{m, n\}$. The complete graph K_n has a matching number equal to $\lfloor n/2 \rfloor$ and covering number equal to $n - 1$. The following theorem of König asserts that for bipartite graphs, the matching and covering numbers are equal.

Theorem 1.1.9 *Let G be a bipartite graph. Then $m(G) = c(G)$, that is, the largest number of edges in a matching equals the smallest number of vertices in a vertex-cover.* □

The notion of isomorphism of graphs is meant to make precise the statement that two graphs are structurally the same.

Definition 1.1.10 Let G be a graph with vertex set V and let H be a graph with vertex set W. An *isomorphism* from G to H is a bijection $\phi : V \to W$ such that $\{x, y\}$ is an edge of G if and only if $\{\phi(x), \phi(y)\}$ is an edge of H. If ϕ is an isomorphism from G to H, then clearly $\phi^{-1} : W \to V$ is an isomorphism from H

to G. The graphs G and H are *isomorphic* provided there is an isomorphism from G to H (and thus one from H to G). The notion of isomorphism carries over to multigraphs by requiring that the edge $\{x, y\}$ occur as many times in G as the edge $\{\phi(x), \phi(y)\}$ occurs in H. □

1.2 Digraphs

In a graph, edges are unordered pairs of vertices and thus have no direction. In a directed graph, edges are ordered pairs of vertices and thus have a direction (or orientation) from the first vertex to the second vertex in the ordered pair. Most of the ideas introduced for graphs can be carried over to directed graphs, modified only to take into account the directions of the edges. As a result, we shall be somewhat brief.

Definition 1.2.1 A *directed graph* (for short, a *digraph*) G consists of a finite set V of elements called *vertices* and a set E of ordered pairs of vertices called (directed) *edges*. The *order* of the digraph G is the number $|V|$ of its vertices. If $\alpha = (x, y)$ is an edge, then x is the *initial vertex* of α and y is the *terminal vertex*, and we say that α is an *edge from x to y*. In case $x = y$, α is a *loop* with initial and terminal vertices both equal to x. A *multidigraph* differs from a digraph in that there may be several edges with the same initial vertex and the same terminal vertex. A *weighted digraph* is a digraph in which each edge has an assigned weight. □

The notions of subgraph, spanning subgraph, and induced subgraph of a graph carry over in the obvious way to *subdigraph*, *spanning subdigraph*, and *induced subdigraph* of a digraph. Digraphs are pictured as graphs, except now the edges have arrows on them to indicate their direction. A digraph G with a spanning subdigraph H_1 and an induced subdigraph H_2 are pictured in Figure 1.5.

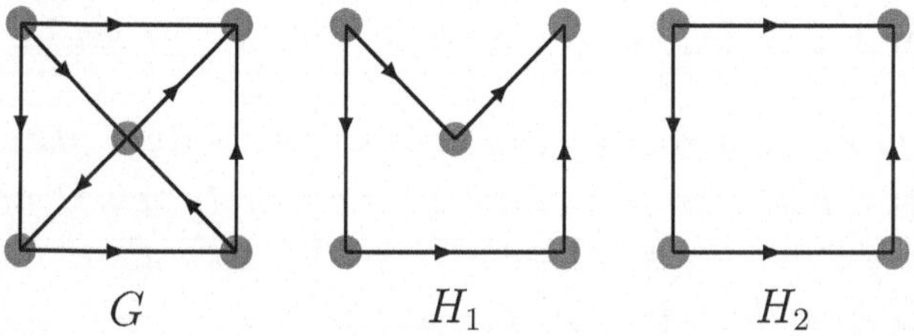

Figure 1.5

In a digraph G, a vertex has two degrees. The *outdegree* $d^+(v)$ of a vertex v is the number of edges of which v is an initial vertex; the *indegree* $d^-(v)$ of v is the number of edges of which v is a terminal vertex. A loop at a vertex contributes 1 to both its indegree and its outdegree. Clearly, the sum of the indegrees of the vertices of a digraph equals the sum of the outdegrees.

Definition 1.2.2 Let G be a digraph. A *walk* in G from vertex u to vertex v is a sequence γ of vertices $u = x_0, x_1, \ldots, x_{k-1}, x_k = v$ such that (x_i, x_{i+1}) is an edge for each $i = 0, 1, \ldots, k-1$. The *edges of the walk* γ are these k edges and γ has *length* k. In a *closed walk*, $u = v$. In a *path*, the vertices $x_0, x_1, \ldots, x_{k-1}, x_k$ are distinct. If $u = v$ and the vertices $x_0, x_1, \ldots, x_{k-1}, x_k$ are otherwise distinct, then the subdigraph consisting of the vertices and edges of γ is a *cycle*. The digraph G is *acyclic* provided it has no cycles. If there is a walk from vertex u to vertex v, then there is a path from u to v. The digraph G is *strongly connected* provided that for each pair u and v of distinct vertices, there is a path from u to v and a path from v to u.

Define $u = v$ provided there is a walk from u to v and a walk from v to u. This is an equivalence relation and thus V is partitioned into equivalence classes $V_1, V_2, \ldots, V_l$. The l subdigraphs induced on the sets of vertices $V_1, V_2, \ldots, V_l$ are the *strong components* of D. The digraph D is strongly connected if and only if it has exactly one strong component. □

The following theorem summarizes some important properties concerning these notions:

Theorem 1.2.3 *Let G be a digraph with vertex set V.*

(i) *Then G is strongly connected if and only if there does not exist a partition of V into two nonempty sets U and W such that all the edges between U and W have their initial vertex in U and their terminal vertex in W.*

(ii) *The strong components of G can be ordered as $G_1, G_2, \ldots, G_l$ so that if (x, y) is an edge of G with x in G_i and y in G_j with $i \neq j$, then $i < j$ (in the ordering $G_1, G_2, \ldots, G_l$ all edges between the strong components go from left to right).* □

Let G be a digraph (or multidigraph) with vertex set V. By replacing each directed edge (x, y) of G by an undirected edge $\{x, y\}$ and deleting any duplicate edges, we obtain a graph G' called the *underlying graph* of G. The digraph G is called *weakly connected* provided its underlying graph G' is connected. The digraph G is called *unilaterally connected* provided that for each pair of distinct vertices u and v, there is a path from u to v or a path from v to u. A unilaterally connected digraph is clearly weakly connected.

The notion of isomorphism of digraphs (and multidigraphs) is quite analogous to that of graphs. The only difference is that the direction of edges has to be taken into account.

Definition 1.2.4 Let G be a digraph with vertex set V, and let H be a digraph with vertex set W. An *isomorphism* from G to H is a bijection $\phi : V \to W$ such that (x, y) is an edge of G if and only if $(\phi(x), \phi(y))$ is an edge of H. If ϕ is an isomorphism from G to H, then $\phi^{-1} : W \to V$ is an isomorphism from H to G. The digraphs G and H are *isomorphic* provided there is an isomorphism from G to H (and thus one from H to G). □

1.3 Some Classical Combinatorics

In this section we review the notions of permutations and combinations and corresponding basic counting formulas.

Definition 1.3.1 Let X be a set with n elements that, for ease of description, we can assume to be the set $\{1, 2, \ldots, n\}$ consisting of the first n positive integers. A *permutation* of X is a listing $i_1 i_2 \ldots i_n$ of the elements of X in some order. There are $n! = n(n-1)(n-2)\cdots 1$ permutations of X. The permutation $i_1 i_2 \ldots i_n$ can be regarded as a bijection $\sigma : X \to X$ from X to X by defining $\sigma(k) = i_k$ for $k = 1, 2, \ldots, n$.

Now let r be a nonnegative integer with $1 \leq r \leq n$. An *r-permutation* of X is a listing $i_1 i_2 \ldots i_r$ of r of the elements of X in some order. There are $n(n-1)\cdots(n-r+1)$ r-permutations of X, and this number can be written as $n!/(n-r)!$. (Here we adopt the convention that $0! = 1$ to allow for the case that $r = n$ in the formula.)

An *r-combination* of X is a selection of r of the objects of X without regard for order. Thus an r-combination of X is just a subset of X with r elements. Each r-combination can be ordered in $r!$ ways, and in this way we obtain all the r-permutations of X. Thus the number of r-combinations of X equals

$$\frac{n!}{r!(n-r)!},$$

a number we denote by $\binom{n}{r}$ (read as *n choose r*). □

For instance,

$$\binom{n}{0} = 1, \binom{n}{1} = n, \binom{n}{2} = \frac{n(n-1}{2}, \ldots, \binom{n}{n-1} = n, \binom{n}{n} = 1.$$

In general, we have

$$\binom{n}{r} = \binom{n}{n-r}, \quad (0 \leq r \leq n),$$

since the complement of an r-combination is a $(n-r)$-combination. The number of combinations (of any size) of the set $\{1, 2, \ldots, n\}$ equals 2^n, since each integer in the set can be chosen or left out of a combination. Counting combinations by size $k = 0, 1, 2, \ldots, n$, we thus get the identity

$$\sum_{k=0}^{n} \binom{n}{k} = 2^n.$$

The above formulas hold for permutations and combinations in which one is not allowed to repeat an object. If we are allowed to repeat objects in a permutation, then more general formulas hold. The number of permutations of $X = \{1, 2, \ldots, n\}$ in which, for each $k = 1, 2, \ldots, n$, the integer k appears m_k times equals

$$\frac{(m_1 + m_2 + \cdots + m_n)!}{m_1!m_2! \cdots m_n!}.$$

This follows by observing that such a permutation is a list of length $N = m_1 + m_2 + \cdots + m_n$, and to form such a list we choose m_1 places for the 1's, m_2 of the remaining places for the 2's, m_3 of the remaining places for the 3's, and so forth, giving

$$\binom{N}{m_1}\binom{N - m_1}{m_2}\binom{N - m_1 - m_2}{m_3} \cdots.$$

After substitution and cancellation, this reduces to the given formula. The number of r-permutations $i_1 i_2 \ldots i_r$ of $X = \{1, 2, \ldots, n\}$, where the number of times each integer in X can be repeated is not restricted, that is, can occur any number of times (sometimes called an *r-permutation-with-repetition*), of X is n^r, since there are n choices for each of the r integers i_k.

For r-combinations of $X = \{1, 2, \ldots, n\}$ in which the number of times an integer occurs is not restricted (other than by the size r of the combination), we have to choose how many times (denote it by x_k) each integer k occurs in the r-combination. Thus the number of such r-combinations equals the number of solutions in nonnegative integers of the equation

$$x_1 + x_2 + \cdots + x_n = r.$$

This is the same as the number of permutations of the two integers 0 and 1 in which 1 occurs r times and 0 occurs $n - 1$ times (the number of 1's to the left of the first 0, in between the 0's, and to the right of the last 0 give the values of $x_1, x_2, \ldots, x_n$). Thus the number of such r-combinations equals

$$\frac{(n+r-1)!}{r!(n-1)!} = \binom{n+r-1}{r} = \binom{n+r-1}{n-1}.$$

Another useful counting technique is provided by the *inclusion-exclusion formula*. Let $X_1, X_2, \ldots, X_n$ be subsets of a finite set U. Then the number of elements of U in none of the sets $X_1, X_2, \ldots, X_n$ is given by

$$|\overline{X_1} \cap \overline{X_2} \cap \cdots \cap \overline{X_n}| = \sum_{k=0}^{n} (-1)^{|K|} \sum_{K \subset \{1,2,\ldots,n\}; |K|=k} |\cap_{i \in K} X_i|.$$

Here $\overline{X_i}$ is the *complement* of X_i in U, that is, the subset of elements of U that are not in X_i. For the value $k = 0$ in the formula, we have $K = \emptyset$, and $\cap_{i \in \emptyset} X_i$ is an intersection over an empty set and is interpreted as U.

The set of $n!$ permutations of $\{1, 2, \ldots, n\}$ can be naturally partitioned into two sets of the same cardinality using properties called evenness and oddness. These properties and the resulting partition are discussed in Chapter 4.

1.4 Fields

The number systems with which we work in this book are primarily the real number system $\Re$ and the complex number system $\mathcal{C}$. But much of what we develop does not use any special properties of these familiar number systems,[1] and works for any number system called a field. We give a working definition of a field since it is not in our interest to systematically develop properties of fields.

[1]One notable exception is that polynomials of degree at least 1 with complex coefficients (in particular, polynomials with real coefficients) always have roots (real or complex). In fact a polynomial of degree $n \geq 1$ with complex coefficients can be completely factored in the form $c(x-r_1)(x-r_2)\cdots(x-r_n)$, where $c, r_1, r_2, \ldots, r_n$ are complex numbers. This property of complex numbers is expressed by saying that the complex numbers are *algebraicaly closed.*

Definition 1.4.1 Let F be a set on which two binary operations[2] are defined, called addition and multiplication, respectively, and denoted as usual by "$+$" and "$\cdot$". Then F is a *field* provided the following properites hold:

(i) (associative law for addition) $a + (b + c) = (a + b) + c$.

(ii) (commutative law for addition) $a + b = b + a$.

(iii) (zero element) There is an element 0 in F such that $a + 0 = 0 + a = a$.

(iv) (additive inverses) Corresponding to each element a, there is an element a' in F such that $a + a' = a' + a = 0$. The element a' is usually denoted by $-a$. Thus $a + (-a) = (-a) + a = 0$.

(v) (associative law for multiplication) $a \cdot (b \cdot c) = (a \cdot b) \cdot c$.

(vi) (commutative law for multiplication) $a \cdot b = b \cdot a$.

(vii) (identity element) There exists an element 1 in F diferent from 0 such that $1 \cdot a = a \cdot 1 = a$.

(viii) (multiplicative inverses) Corresponding to each element $a \neq 0$, there is an element a'' in F such that $a \cdot a'' = a'' \cdot a = 1$. The element a'' is usually denoted by a^{-1}. Thus $a \cdot a^{-1} = a^{-1} \cdot a = 1$.

(ix) (distributive laws) $a \cdot (b + c) = a \cdot b + a \cdot c$ and $(b + c) \cdot a = b \cdot a + c \cdot a$.

It is understood that the above properties are to hold for all choices of the elements a, b, and c in F. Note that properties (i)–(iv) involve only addition and properties (v)–(viii) involve only multiplication. The distributive laws connect the two binary operations and make them dependent on one another. We often drop the multiplication symbol and write ab in place of $a \cdot b$. Thus, for instance, the associative law (v) becomes $a(bc) = (ab)c$. □

---

[2] A binary operation on F means that given an ordered pair a, b of elements in F, they can be combined using the operation to produce another element in F. This is sometimes expressed by saying that the operation of combining two elements satisfies the *closure property*.

Examples of fields are (a) the set $\Re$ of real numbers with the usual addition and multiplication, (b) the set $\mathcal{C}$ of complex numbers with the usual addition and multiplication, and (c) the set $\mathcal{Q}$ of rational numbers with the usual addition and multiplication. A familiar number systen that is *not* a field is the set of integers with the usual addition and multiplication (e.g., 2 does not have a multiplicative inverse).

Properties (i), (iii), and (iv) are the defining properties for an algebraic system with one binary operation, denoted here by $+$, called a *group*. If property (ii) also holds then we have a *commutative group*. By properties (v)–(viii) the nonzero elements of a field form a commutative group under the binary operation of multiplication.

In the next theorem we collect a number of elementary properties of fields whose proofs are straightforward.

Theorem 1.4.2 *Let F be a field. Then the following hold:*

(i) *The zero element* 0 *and identity element* 1 *are unique.*

(ii) *The additive inverse of an element of F is unique.*

(iii) *The multiplicative inverse of a nonzero element of F is unique.*

(iv) $a \cdot 0 = 0 \cdot a = 0$ *for all a in F.*

(v) $-(-a) = a$ *for all a in F.*

(vi) $(a^{-1})^{-1} = a$ *for all nonzero a in F.*

(vii) *(cancellation laws) If $a \cdot b = 0$, then $a = 0$ or $b = 0$. If $a \cdot b = a \cdot c$ and $a \neq 0$, then $b = c$.* □

We now show how one can construct fields with a finite number of elements. Let m be a positive integer. First we recall the *division algorithm*, which asserts that if a is any integer, there are unique integers q (the *quotient*) and r (the *remainder*), with $0 \leq r \leq m-1$, such that $a = qm+r$. For integers a and b, define a to be *congruent modulo m* to b, denoted $a \equiv b \pmod{m}$, provided

m is a divisor of $a - b$. Congruence modulo m is an equivalence relation, and as a result the set Z of integers is partitioned into equivalence classes. The equivalence class containing a is denoted by $[a]_m$. Thus $[a]_m = [b]_m$ if and only if m is a divisor of $a - b$.

It follows easily that $a \equiv b \pmod{m}$ if and only if a and b have the same remainder when divided by m. Thus there is a one-to-one correspondence between equivalence classes modulo m and the possible remainders $0, 1, 2, \ldots, m-1$ when an integer is divided by m. We can thus identify the equivalence classes with $0, 1, 2, \ldots, m-1$. Congruence satisfies a basic property with regard to addition and mutltiplication that is easily verified:

If $a \equiv b \pmod{m}$ and $c \equiv d \pmod{m}$, then

$$a + c \equiv b + d \pmod{m} \text{ and } ac \equiv bd \pmod{m}.$$

This property allows one to add and multiply equivalence classes unambiguously as follows:

$$[a]_m + [b]_m = [a+b]_m \text{ and } [a]_m \cdot [b]_m = [ab]_m.$$

Let $Z_m = \{0, 1, 2, \ldots, m-1\}$. Then Z_m contains exactly one element from each equivalence class, and we can regard addition and multiplication of equivalence classes as addition and multiplication of integers in Z_m. For instance, let $m = 9$. Then, examples of addition and multiplication in Z_9 are

$$\begin{aligned} 4 + 3 = 7 &\quad \text{and } 6 + 7 = 4 \\ 5 + 0 = 5 &\quad \text{and } 1 \cdot 6 = 6 \\ 4 \cdot 8 = 5 &\quad \text{and } 7 \cdot 4 = 1 \end{aligned}$$

If m is a prime number, then, as shown in the next theorem, Z_m is actually a field. To prove this, we require another basic property of integers, namely, that if a and m are integers whose greatest common divisor is d, then there are integers s and t expressing d as a linear integer combination of a and m:

$$d = sa + tm.$$

Theorem 1.4.3 *Let m be a prime number. With the addition and multiplication as defined above, Z_m is a field.*

Proof. Most of the proof of this theorem is routine. It is clear that $0 \in Z_m$ and $1 \in Z_m$ are the zero element and identity element. If $a \in Z_m$ and $a \neq 0$, then $m - a$ is the additive inverse of a. If $a \in Z_m$ and $a \neq 0$, then the greatest common divisor of a and m is 1, and hence there exist integers s and t such that $sa + tm = 1$. Thus $sa = 1 - tm$ is congruent to 1 modulo m. Let s^* be the integer in Z_m congruent to s modulo m. Then we also have $s^*a \equiv 1 \bmod m$. Hence s^* is the multiplicative inverse of a modulo m. Verification of the rest of the field properties is now routine. □

As an example, let $m = 7$. Then Z_7 is a field with

$$\begin{array}{ll} 2 \cdot 4 = 1 & \text{so that } 2^{-1} = 4 \text{ and } 4^{-1} = 2; \\ 3 \cdot 5 = 1 & \text{so that } 3^{-1} = 5 \text{ and } 5^{-1} = 3; \\ 6 \cdot 6 = 1 & \text{so that } 6^{-1} = 6. \end{array}$$

Two fields F and F' are *isomorphic* provided there is a bijection $\phi : F \to F'$ that preserves both addition and multiplication:

$$\phi(a + b) = \phi(a) + \phi(b), \text{ and } \phi(a \cdot b) = \phi(a) \cdot \phi(b).$$

In these equations the leftmost binary operations (addition and multiplication, respectively) are those of F and the rightmost are those of F'. It is a fundamental fact that any two fields with the same finite number of elements are isomorphic.

1.5 Vector Spaces

There is an important, abstract notion of a vector space over a field that does not have to concern us here. We shall confine our attention to the *vector space F^n of n-tuples over a field F*, whose elements are called *vectors*, that is,

$$F^n = \{(a_1, a_2, \ldots, a_n) : a_i \in F, i = 1, 2, \ldots, n\}.$$

The zero vector is the n-tuple $(0, 0, \ldots, 0)$, where 0 is the zero element of F. As usual, the zero vector is also denoted by 0 with the context determining whether the zero element of F or the zero vector is intended. The elements of F are now called *scalars.*

Using the addition and multiplication of the field F, vectors can be added componentwise and multiplied by scalars. Let $u = (a_1, a_2, \ldots, a_n)$ and $v = (b_1, b_2, \ldots, b_n)$ be in F^n. Then

$$u + v = (a_1 + b_1, a_2 + b_2, \ldots, a_n + b_n).$$

If c is in F, then

$$cu = (ca_1, ca_2, \ldots, ca_n).$$

Since vector addition and scalar multiplication are defined in terms of addition and multiplication in F that satisfy certain associative, commutative, and distributive laws, we obtain associative, commutative, and distributive laws for vector addition and scalar multiplication. These laws are quite transparent from those for F, and we only mention the following:

(i) $u + 0 = 0 + u = u$ for all vectors u.

(ii) $0u = u0 = 0$ for all vectors u.

(iii) $u + v = v + u$ for all vectors u and v.

(iv) $(c + d)u = cu + du$ for all vectors u and scalars c and d.

(v) $c(u + v) = cu + cv$ for all vectors u and v and scalars c.

(vi) $1u = u$ for all vectors u.

(vii) $(-1)u = (-u_1, -u_2, \ldots, -u_n)$ for all vectors u; this vector is denoted by $-u$ and is called the *negative* of u and satisfies $u + (-u) = (-u) + u = 0$, for all vectors u.

(viii) $c(du) = (cd)u$ for all vectors u and scalars c and d.

A fundamental notion is that of a subspace of F^n. Let V be a nonempty subset of F^n. Then V is a *subspace* of F^n provided V is closed under vector addition and scalar multiplication, that is,

(a) For all u and v in V, $u + v$ is also in V.

(b) For all u in V and c in F, cu is in V.

Let u be in the subspace V. Because $0u = 0$, it follows that the zero vector is in V. Similarly, $-u$ is in V for all u in V. A simple example of a subspace of F^n is the set of all vectors $(0, a_2, \ldots, a_n)$ with first coordinate equal to 0. The zero vector itself is a subspace.

Definition 1.5.1 Let $u^{(1)}, u^{(2)}, \ldots, u^{(m)}$ be vectors in F^n, and let $c_1, c_2, \ldots, c_m$ be scalars. Then the vector

$$c_1 u^{(1)} + c_2 u^{(2)} + \cdots + c_m u^{(m)}$$

is called a *linear combination* of $u^{(1)}, u^{(2)}, \ldots, u^{(m)}$. If V is a subspace of F^n, then V is closed under vector addition and scalar multiplication, and it follows easily by induction that a linear combination of vectors in V is also a vector in V. Thus *subspaces are closed under linear combinations*; in fact, this can be taken as the defining property of subspaces. The vectors $u^{(1)}, u^{(2)}, \ldots, u^{(m)}$ *span* V (equivalently, form a *spanning set* of V) provided every vector in V is a linear combination of $u^{(1)}, u^{(2)}, \ldots, u^{(m)}$. The zero vector can be written as a linear combination of $u^{(1)}, u^{(2)}, \ldots, u^{(m)}$ with all scalars equal to 0; this is a *trivial linear combination*. The vectors $u^{(1)}, u^{(2)}, \ldots, u^{(m)}$ are *linearly dependent* provided there are scalars $c_1, c_2, \ldots, c_m$, not all of which are zero, such that

$$c_1 u^{(1)} + c_2 u^{(2)} + \cdots + c_m u^{(m)} = 0,$$

that is, the zero vector can be written as a *nontrivial linear combination* of $u^{(1)}, u^{(2)}, \ldots, u^{(m)}$. For example, the vectors $(1, 4), (3, -1)$, and $(3, 5)$ in $\Re^2$ are linearly dependent since

$$3(1, 4) + 1(3, -2) - 2(3, 5) = (0, 0).$$

Vectors are *linearly independent* provided they are not linearly dependent. The vectors $u^{(1)}, u^{(2)}, \ldots, u^{(m)}$ are a *basis* of V provided they are linearly independent and span V. By an *ordered basis* we mean a basis in which the vectors of the basis are listed in a specified order; to indicate that we have an ordered basis we write $(u^{(1)}, u^{(2)}, \ldots, u^{(m)})$. A spanning set S of V is a *minimal spanning set of* V provided that each set of vectors obtained from S by removing a vector is not a spanning set for V. A linearly independent set S of vectors of V is a *maximal linearly independent set of vectors of* V provided that for each vector w of V that is not in S, $S \cup \{w\}$ is linearly dependent (when this happens, w must be a linear combination of the vectors in S). □

In the next theorem, we collect some basic facts about these properties.

Theorem 1.5.2 *Let V be a subspace of F^n.*

(i) *Then V has a basis and any two bases of V contain the same number of vectors.*

(ii) *A minimal spanning set of V is a basis of V. Thus every spanning set of vectors contains a basis of V.*

(iii) *A maximal linearly independent set of vectors of V is a basis of V. Thus every linearly independent set of vectors can be extended to a basis of V.*

(iv) *If $(u^{(1)}, u^{(2)}, \ldots, u^{(m)})$ is an ordered basis of V, then each vector u in V can be written uniquely as a linear combination of these vectors: $u = c_1u^{(1)} + c_2u^{(2)} + \cdots + c_mu^{(m)}$, where the scalars $c_1, c_2, \ldots, c_m$ are uniquely determined.* □

The number of vectors in a basis of a subspace V and so, by (i) of Theorem 1.5.2, the number of vectors in every basis of V, is the *dimension* of V, denoted by $\dim V$. In (iv) of Theorem 1.5.2, the scalars $c_1, c_2, \ldots, c_n$ are the *coordinates* of u with respect to the ordered basis $(u^{(1)}, u^{(2)}, \ldots, u^{(m)})$.

Definition 1.5.3 Let U be a subspace of F^m and let V be a subspace of F^n. A mapping $T : U \to V$ is a *linear transformation* provided

$$T(cu + dw) = cT(u) + dT(v)$$

for all vectors u and w in U and all scalars c and d. The *kernel* of the linear transformation T is the set

$$\ker(T) = \{u \in U : T(u) = 0\}$$

of all vectors of U that are mapped to the zero vector of V. The linear transformation T is an *injective linear transformation* if and only if $\ker(T) = \{0\}$. The *range* of T is the set

$$\text{range}(T) = \{T(u) : u \in U\}$$

of all values (images) of vectors in U. It follows by induction from the definition of a linear transformation that linear transformations preserve all linear combinations, that is,

$$T(c_1u^{(1)}+c_2u^{(2)}+\cdots+c_ku^{(k)}) = c_1T(u^{(1)})+c_2T(u^{(2)})+\cdots+T(c_ku^{(k)})$$

for all vectors $u^{(1)}, u^{(2)}, \ldots, u^{(k)}$ and all scalars $c_1, c_2, \ldots, c_k$. □

Finally, we review the notion of the dot product of vectors in $\Re^n$ and $\mathcal{C}^n$.

Definition 1.5.4 Let $u = (a_1, a_2, \ldots, a_n)$ and $v = (b_1, b_2, \ldots, b_n)$ be vectors in either $\Re_n$ or $\mathcal{C}^n$. Then their *dot product* $u \cdot v$ is defined by

(i) $u \cdot v = a_1b_1 + a_2b_2 + \cdots a_nb_n, \; u, v \in \Re^n$;

(ii) $u \cdot v = a_1\overline{b_1} + a_2\overline{b_2} + \cdots a_n\overline{b_n}, \; u, v \in \mathcal{C}^n$.

Here $\overline{b}$ denotes the complex conjugate[3] of b. In particular, we have that

$$u \cdot u = a_1a_1 + a_2\overline{a_2} + \cdots + a_n\overline{a_n} = |a_1|^2| + |a_2|^2 + \cdots + |a_n|^2 \geq 0$$

with equality if and only if u is a zero vector. The *norm* (or *length*) $||u||$ of a vector u is defined by

$$||u|| = \sqrt{u \cdot u}.$$

□

[3]Recall that $\overline{a+b} = \overline{a} + \overline{b}$ and $\overline{ab} = \overline{a}\overline{b}$.

The next theorem contains some elementary properties of dot products and norms.

Theorem 1.5.5 *Let u, v, and w be vectors in $\Re^n$ or C^n. Then the following hold:*

(i) $||u|| \geq 0$ *with equality if and only if* $u = 0$.

(ii) $(u + v) \cdot w = u \cdot w + v \cdot w$ *and* $u \cdot (v + w) = u \cdot v + u \cdot w$.

(iii) $cu \cdot v = c(u \cdot v)$ *and* $u \cdot cv = \bar{c}u \cdot v$ *(so if c is a real scalar,* $u \cdot cv = c(u \cdot v)$*).*

(iv) $u \cdot v = \overline{v \cdot u}$ *(so if u and v are real vectors.* $u \cdot v = v \cdot u$*)*

(v) *(Cauchy–Schwarz inequality)* $|u \cdot v| \leq ||u|| ||v||$ *with equality if and only if u and v are linearly dependent.*

□

Let u and v be nonzero vectors in $\Re^n$. By the Cauchy–Schwarz inequality,

$$-1 \leq \frac{u \cdot v}{||u|| ||v|} \leq 1.$$

Hence there is an angle θ with $0 \leq \theta \leq \pi$ such that

$$\cos\theta = \frac{u \cdot v}{||u|| ||v|}.$$

The angle θ is the *angle between the vectors u and v.* It follows that $u \cdot v = 0$ if and only if $\theta = \pi/2$, in which case u and v are *orthogonal.* The zero vector is orthogonal to every vector. For vectors u and v in C^n, we also say that u and v are orthogonal if $u \cdot v = 0$. Mutually orthogonal, nonzero vectors are linearly independent. In particular, n mutually orthogonal, nonzero vectors $u_1, u_2, \ldots, u_n$ of $\Re^n$ or C^n form a basis. If, in addition, each of the vectors $u_1, u_2, \ldots, u_n$ has unit length (which can always be achieved by multiplying each u_i by $1/||u_i||$), then $u_1, u_2, \ldots, u_n$ is an *orthonormal basis.*

Now let $v_1, v_2, \ldots, v_m$ be an arbitrary basis of a subspace V of $\Re^n$ or of C^n. The *Gram–Schmidt orthogonalization algorithm* determines an orthonormal basis $u_1, u_2, \ldots, u_m$ with the property that the subspace spanned by $v_1, v_2, \ldots, v_k$ equals the subspace spanned by $u_1, u_2, \ldots, u_k$ for each $k = 1, 2, \ldots, m$. After first normalizing v_1 to obtain a vector $u_1 = v_1/||v_1||$ of unit length, the algorithm proceeds recursively by orthogonally projecting v_{i+1} onto the subspace V_i spanned by $u_1, \ldots, u_i$ (equivalently, the subspace spanned by $v_1, \ldots, v_i$), forming the difference vector that is orthogonal to this subspace V_i, and then normalizing this vector to have length 1. Algebraically, we have

$$u_{i+1} = \frac{v_{i+1} - \text{Proj}_{V_i}(v_{i+1})}{||v_{i+1} - \text{Proj}_{V_i}(v_{i+1})||} = \frac{v_{i+1} - \sum_{j=1}^{i}(v_j \cdot u_j)u_j}{||v_{i+1} - \sum_{j=1}^{i}(v_j \cdot u_j)u_j||},$$

for $i = 1, 2, \ldots, m-1$.

For each θ with $0 \leq \theta \leq \pi$, the vector $(\cos\theta, \sin\theta)^T$ and the vector $(-\sin\theta, \cos\theta)^T$ form an orthonormal basis of $\Re^2$; this is the orthonormal basis obtained by rotating the standard basis $(1, 0), (0, 1)$ by an angle θ in the counterclockwise direction.

In this first chapter, we have given a very brief introduction to elementary graph theory, combinatorics, and linear algebra. For more about these areas of mathematics, and indeed for many of the topics discussed in the next chapters, one may consult the extensive material given in the handbooks *Handbook of Discrete and Combinatorial Mathematics* [68], *Handbook of Graph Theory* [39], and *Handbook of Linear Algebra* [46].

1.6 Exercises

1. Prove Theorem 1.1.4.

2. List the structurally different trees of order 6.

3. Prove that there does not exist a regular graph of degree k with n vertices if both n and k are odd.

4. Determine the chromatic numbers of the following graphs: (a) the graph obtained from K_n by removing an edge; (b) the graph obtained from K_n by removing two edges (there are

two possibilities: the removed edges may or may not have a vertex in common); (c) the graph obtained from a tree by adding a new edge (the new edge may create either a cycle of even length or a cycle of odd length).

5. Let G be the bipartite graph with bipartition

$$U = \{u_1, u_2, u_3, u_4, u_5, u_6\} \text{ and } W = \{w_1, w_2, w_3, w_4, w_5, w_6\}$$

whose edges are all those pairs $\{u_i, w_j\}$ for which $2i + 3j$ is congruent to 0, 1, or 5 modulo 6. Draw the graph G and determine a matching with the largest number of edges and a vertex-cover with the smallest number of vertices.

6. Let the digraph G be obtained from the complete graph K_n by giving a direction to each edge. (Such a digraph is usually called a *tournament.*) Let $d_1^+, d_2^+, \ldots, d_n^+$ be the outdegrees of G in some order. Prove that

$$d_1^+ + d_2^+ + \cdots + d_k^+ \leq \binom{k}{2} \quad (k = 1, 2, \ldots, n)$$

with equality for $k = n$.

7. Let D be the digraph with vertex set $\{1, 2, 3, 4, 5, 6, 7, 8\}$, where there is an edge from i to j if and only if $2i + 3j$ is congruent to 1 or 4 modulo 8. Determine whether or not D is strongly connected.

8. Use the inclusion-exclusion formula to show that the number of permutations $i_1 i_2 \ldots i_n$ of $\{1, 2, \ldots, n\}$ such that $i_k \neq k$ for $k = 1, 2, \ldots, n$ is given by

$$n! \sum_{j=0}^{n} \frac{(-1)^k}{k!}.$$

9. Prove that the number of even (respectively, odd) combinations of $\{1, 2, \ldots, n\}$ equals 2^{n-1}. (By an even combination we understand a combination with an even number of elements; an odd combination has an odd number of elements.)

10. Determine the number of solutions in nonnegative integers of

$$x_1 + x_2 + x_3 + x_4 + x_5 = 24,$$

where $x_1 \geq 2$ and $x_5 \geq 3$.

11. Write out the complete addition and multiplication tables for the field Z_5.

12. Prove Theorem 1.4.2.

13. Show that $101^{500} \equiv 1 \pmod{100}$ and that $99^{501} \equiv -1 \pmod{100}$.

14. Let V be the set of all vectors $(a_1, a_2, \ldots, a_n)$ in F^n such that $a_1 + a_2 + \ldots + a_n = 0$. Prove that V is a subspace of F^n and find a basis of V.

15. Let $u^{(1)}, u^{(2)}, \ldots, u^{(n)}$ be an orthonormal basis of $\Re^n$. Prove that if u is a vector in $\Re^n$, then

$$u = \sum_{i=1}^{n} (u \cdot u^{(i)}) u^{(i)}.$$

16. Prove Theorem 1.5.5.

17. Show that $(1, 0, 0), (1, 1, 0), (1, 1, 1)$ is a basis of $\Re^3$ and use the Gram–Schmidt orthogonalization algorithm to obtain an orthonormal basis.

Chapter 2

Basic Matrix Operations

In this chapter we introduce matrices as arrays of numbers and define their basic algebraic operations: sum, product, and transposition. Next, we associate to a matrix a digraph called the König digraph and establish connections of matrix operations with certain operations on graphs. These graph-theoretic operations illuminate the matrix operations and aid in understanding their properties. In particular, we use the König digraph to explain how matrices can be partitioned into blocks in order to facilitate matrix operations.

2.1 Basic Concepts

Let m and n be positive integers.[1] A *matrix* is an m by n rectangular array of numbers[2] of the form

$$A = \begin{bmatrix} a_{11} & a_{12} & \cdots & a_{mn} \\ a_{21} & a_{22} & \cdots & a_{2n} \\ \vdots & \vdots & \ddots & \vdots \\ a_{m1} & a_{m2} & \cdots & a_{mn} \end{bmatrix}. \tag{2.1}$$

[1]There will be occasions later when we will want to allow m and n to be 0, resulting in empty matrices in the definition.

[2]These may be real numbers, complex numbers, or numbers from some other arithmetic system, such as the integers modulo n.

The matrix A has *size* m by n and we often say that its *type* is $m \times n$. The mn numbers a_{ij} are called the *entries* or (*elements*) of the matrix A. If $m = n$, then A is a *square matrix*, and instead of saying A has size n by n we usually say that A is a *square matrix of order* n.

The matrix A in (2.1) has m rows of the form

$$\alpha_i = \begin{bmatrix} a_{i1} & a_{i2} & \cdots & a_{in} \end{bmatrix}, \quad (i = 1, 2, \ldots, m)$$

and n columns

$$\beta_j = \begin{bmatrix} a_{1j} \\ a_{2j} \\ \vdots \\ a_{mj} \end{bmatrix}, \quad (j = 1, 2, \ldots, n).$$

The entry a_{ij} contained in both α_i and β_j, that is, the entry at the intersection of row i and column j, is the (i, j)*-entry* of A. The rows α_i are 1 by n matrices, or *row vectors*; the columns β_j are m by 1 matrices, or *column vectors*. For brevity we denote the m by n matrix A by

$$A = [a_{ij}]_{m,n}$$

and usually more simply as $[a_{ij}]$ if the size is understood.

Two matrices $A = [a_{ij}]$ and $B = [b_{ij}]$ are *equal matrices* provided that they have the same size m by n and corresponding entries are equal:

$$a_{ij} = b_{ij}, \quad (i = 1, 2, \ldots, m; j = 1, 2, \ldots, n).$$

Thus, for instance, a 2 by 3 matrix can never equal a 3 by 2 matrix, and

$$\begin{bmatrix} 2 & 0 & 5 \\ 1 & 3 & 4 \end{bmatrix} \neq \begin{bmatrix} 2 & 1 & 5 \\ 1 & 3 & 4 \end{bmatrix},$$

because the $(1, 2)$-entries of these matrices are not equal.

Addition and subtraction of matrices are defined in a natural way by the addition and subtraction of corresponding entries. More precisely, if $A = [a_{ij}]$ and $B = [b_{ij}]$ are matrices of the same size m by n, then their *matrix sum* is the m by n matrix

$$A + B = [a_{ij} + b_{ij}]$$

and their *matrix difference* is the m by n matrix

$$A - B = [a_{ij} - b_{ij}].$$

In both cases we perform the operation entrywise.

Example 2.1.1

$$\begin{bmatrix} 1 & -3 & 0 \\ 4 & 2 & -3 \end{bmatrix} + \begin{bmatrix} 2 & 3 & 4 \\ -1 & 5 & 6 \end{bmatrix} = \begin{bmatrix} 3 & 0 & 4 \\ 3 & 7 & 3 \end{bmatrix}.$$

Two matrices of different sizes can never be added or subtracted. □

The multiplication of matrices is more complicated and, as a result, more interesting and, as we shall see, important and useful. First of all, the multiplication $A \cdot B$, or simply AB, of two matrices A and B is possible if and only if the number of columns of A equals the number of rows of B. So let $A = [a_{ij}]_{m,n}$ and $B = [b_{ij}]_{n,p}$. Then the *matrix product* $A \cdot B$ (more simply, AB) is the m by p matrix $C = [c_{ij}]$, where

$$c_{ij} = a_{i1}b_{1j} + a_{i2}b_{2j} + \cdots + a_{in}b_{nj}, \quad (i = 1, 2, \ldots, m; j = 1, 2, \ldots, p).$$

Thus the (i, j)-entry of AB is determined only by the ith row of A and the jth column of B.

Example 2.1.2 We have the matrix product

$$\begin{bmatrix} 1 & -3 & 0 \\ 4 & 2 & -3 \end{bmatrix} \begin{bmatrix} 1 & 2 & 5 & -1 \\ 0 & 3 & -2 & 5 \\ 4 & 0 & 1 & 6 \end{bmatrix} = \begin{bmatrix} 1 & -7 & 11 & -16 \\ -8 & 14 & 13 & -12 \end{bmatrix}.$$

Here, for instance, $1 = 4 \cdot 5 + 2 \cdot (-2) + (-3) \cdot 1 = 20 - 4 - 3$. □

There are some important observations to be made here. First, even though the product AB is defined (because the number of columns of A equals the number of rows of B), the product BA may not be defined (because the number of columns of B may not equal the number of rows of A). In fact, if A is m by n, then both AB and BA are defined if and only if B is n by m. In particular, if A and B are square matrices of the same order n, then both AB and BA are defined. But they need not be equal matrices.

Example 2.1.3 We have

$$\begin{bmatrix} 0 & 1 \\ 0 & 0 \end{bmatrix}\begin{bmatrix} 1 & 0 \\ 0 & 0 \end{bmatrix} = \begin{bmatrix} 0 & 0 \\ 0 & 0 \end{bmatrix},$$

while

$$\begin{bmatrix} 1 & 0 \\ 0 & 0 \end{bmatrix}\begin{bmatrix} 0 & 1 \\ 0 & 0 \end{bmatrix} = \begin{bmatrix} 0 & 1 \\ 0 & 0 \end{bmatrix}.$$

Notice also that we have, in this case, an instance of matrix multiplication where $BA = A$. □

In addition to matrix addition, subtraction, and multiplication, there is one additional operation that we define now. It's perhaps the simplest of them all. Let $A = [a_{ij}]$ be an m by n matrix and let c be a number. Then the matrix $c \cdot A$, or simply cA, is the m by n matrix obtained by multiplying each entry of A by c:

$$cA = [ca_{ij}].$$

The matrix cA is called a *scalar multiple* of A.

Matrix transposition is an operation defined on one matrix by interchanging rows with columns in the following way. Let

$$A = \begin{bmatrix} a_{11} & a_{12} & \cdots & a_{1n} \\ a_{21} & a_{22} & \cdots & a_{2n} \\ \vdots & \vdots & \ddots & \vdots \\ a_{m1} & a_{m2} & \cdots & a_{mn} \end{bmatrix}$$

be an m by n matrix. Then the *transpose of the matrix* A is the n by m matrix

$$A^T = \begin{bmatrix} a_{11} & a_{21} & \cdots & a_{m1} \\ a_{12} & a_{22} & \cdots & a_{m2} \\ \vdots & \vdots & \ddots & \vdots \\ a_{1n} & a_{2n} & \cdots & a_{mn} \end{bmatrix}.$$

Matrix transposition satisfies the following properties (where the matrices are assumed to be of the appropriate sizes so that the operations can be carried out), and these properties can be verified in a straightforward manner:

$$\begin{aligned}
(A^T)^T &= A \text{ (transposition is an } \textit{involutory operation}\text{)},\\
(A+B)^T &= A^T + B^T \text{ (transposition commutes with addition)},\\
(cA)^T &= cA^T\\
&\qquad \text{(transposition commutes with scalar multiplication)}.
\end{aligned}$$

Elementary, but not as straightforward, is the relation

$$\begin{aligned}
(AB)^T &= B^T A^T,\\
&\qquad \text{(transposition ``anticommutes'' with multiplication)}.
\end{aligned}$$

This relationship can be verified by observing that the entry in position (i,j) of $(AB)^T$ (so the (j,i)-entry of AB) is obtained from the jth row of A and the ith column of B as prescribed by matrix multiplication, while the entry in position (i,j) of $B^T A^T$ is obtained from the ith row of B^T (so the ith column of B) and the jth column of A^T (so the jth row of A), again as prescribed by matrix multiplication. In the next section, we give a graph-theoretic viewpoint of the relation $(AB)^T = B^T A^T$.

To conclude this section we define some special matrices that are very useful. A *zero matrix* is a matrix each of whose entries equals 0. A zero matrix of size m by n is denoted by $O_{m,n}$. We often simply write O with the size of the matrix being understood from the context. An *identity matrix* (or *unit matrix*) is a square matrix $A = [a_{ij}]$ such that $a_{ii} = 1$ for all i and $a_{ij} = 0$ if $i \neq j$. An identity matrix of order n is denoted by I_n, and we often write I with the order being understood from the context. Thus the identity matrix of order n is the matrix $I_n = [\delta_{ij}]_n$ where δ is the so-called *Kronecker δ-symbol* defined by

$$\delta_{ij} = \begin{cases} 1, & \text{if } i = j,\\ 0, & \text{if } i \neq j. \end{cases}$$

Example 2.1.4 The identity matrix of order 3 is

$$I_3 = \begin{bmatrix} 1 & 0 & 0\\ 0 & 1 & 0\\ 0 & 0 & 1 \end{bmatrix}.$$

□

For matrix addition, zero matrices act like the number 0 acts for ordinary addition. For matrix multiplication, identity matrices act like the number 1 acts for ordinary multiplication. These properties are expressed more precisely in the following equations, which are easily verified:

$$O + A = A + O = A \tag{2.2}$$

if A and O have the same size, and

$$I_m A = A \text{ and } BI_n = B, \tag{2.3}$$

if A has m rows and B has n columns.

The *main diagonal* or simply *diagonal*, of a square matrix $A = [a_{ij}]$ of order n consists of the n entries $a_{11}, a_{22}, \ldots, a_{nn}$. We also refer to the n positions of these n entries of A as the main diagonal of A, and we refer to the remaining positions of A as the *off-diagonal* of A. A square matrix is a *diagonal matrix* provided each off-diagonal entry of A equals 0.

Example 2.1.5 The matrix

$$\begin{bmatrix} 3 & 0 & 0 \\ 0 & -1 & 0 \\ 0 & 0 & 2 \end{bmatrix}$$

is a diagonal matrix. Identity matrices as well as square zero matrices are diagonal matrices. □

A diagonal matrix with diagonal entries $d_1, d_2, \cdots, d_n$ is sometimes denoted by

$$D = \text{diag}(d_1, d_2, \cdots, d_n).$$

If $d_1 = d_2 = \cdots = d_n$, with the common value equal to d, then $D = dI_n$ and D is called a *scalar matrix.* A square matrix is an *upper triangular matrix* provided all its entries below the main diagonal equal zero (thus the nonzero entries are confined to those positions on and above the main diagonal). A *lower triangular matrix* is defined in an analogous way. Note that if A is a square matrix, then A is upper triangular if and only if A^T is lower triangular.

A *permutation matrix* $P = [p_{ij}]$ of order m is a square matrix that has exactly one 1 in each row and column and 0's elsewhere. Thus a permutation matrix of order m has exactly m nonzero entries and each of these m entries equals 1. Permutation matrices correspond to permutations in the following way: Let $\sigma = k_1k_2\ldots k_m$ be a permutation of $\{1, 2, \ldots, m\}$. Let $P = [p_{ij}]$ be the square matrix of order m defined by

$$p_{ij} = \begin{cases} 1, & \text{if } j = k_i, \\ 0, & \text{otherwise.} \end{cases}$$

Then P is a permutation matrix and every permutation matrix of order m corresponds to a permutation of $\{1, 2, \ldots, m\}$ in this way. If A is an m by n matrix, then PA is obtained by permuting the rows of A so that in PA the rows of A are in the order: row k_1, row k_2, ..., row k_m.

Example 2.1.6 If $\sigma = 3124$, then

$$P = \begin{bmatrix} 0 & 0 & 1 & 0 \\ 1 & 0 & 0 & 0 \\ 0 & 1 & 0 & 0 \\ 0 & 0 & 0 & 1 \end{bmatrix}.$$

□

The definition of a permutation matrix treats rows and columns in the same way. Thus the transpose of a permutation matrix is a permutation matrix, and we have from the properties of transposition that

$$(PA)^T = A^T P^T.$$

It thus follows that to permute the columns of an m by n matrix so that they occur in the order $l_1, l_2, \cdots, l_n$, we multiply A on the right by the permutation matrix Q^T, where Q is the permutation matrix corresponding to the permutation $l_1 l_2 \cdots l_n$ of $\{1, 2, \ldots, n\}$. In particular, if A is a square matrix of order n, then the matrix QAQ^T is obtained from A by permuting the rows to put them in the order $l_1, l_2, \ldots, l_n$ and permuting the columns to put them in the order $l_1, l_2, \ldots, l_n$. The matrix QAQ^T is obtained from A by *simultaneous permutations* of its rows and columns.

Example 2.1.7 Let $A = [a_{ij}]$ be a general matrix of order 4. Let P be the permutation matrix corresponding to the permutation 2431 of $\{1, 2, 3, 4\}$. Then

$$\begin{aligned} PAP^T &= \begin{bmatrix} 0 & 1 & 0 & 0 \\ 0 & 0 & 0 & 1 \\ 0 & 0 & 1 & 0 \\ 1 & 0 & 0 & 0 \end{bmatrix} \begin{bmatrix} a_{11} & a_{12} & a_{13} & a_{14} \\ a_{21} & a_{22} & a_{23} & a_{24} \\ a_{31} & a_{32} & a_{33} & a_{34} \\ a_{41} & a_{42} & a_{43} & a_{44} \end{bmatrix} \begin{bmatrix} 0 & 0 & 0 & 1 \\ 1 & 0 & 0 & 0 \\ 0 & 0 & 1 & 0 \\ 0 & 1 & 0 & 0 \end{bmatrix} \\ &= \begin{bmatrix} a_{22} & a_{24} & a_{23} & a_{21} \\ a_{42} & a_{44} & a_{43} & a_{41} \\ a_{32} & a_{34} & a_{33} & a_{31} \\ a_{12} & a_{14} & a_{13} & a_{11} \end{bmatrix}. \end{aligned}$$

Note that the main diagonal entries of A occur in the order 2, 4, 3, 1 on the main diagonal of PAP^T. □

Finally, let A and B be matrices of sizes m by n and p by q, respectively. Then the *direct sum* of A with B is the $m + p$ by $n + q$ matrix given by

$$A \oplus B = \begin{bmatrix} A & O_{m,q} \\ O_{p,n} & B \end{bmatrix}.$$

In case A and B are square matrices, so is their direct sum. The direct sum of more than two matrices is defined in the obvious way.

In the next section, we introduce the König digraph of a matrix that illuminates much of our discussion in this section.

2.2 The König Digraph of a Matrix

Let $A = [a_{ij}]$ be an m by n matrix. Corresponding to A we introduce a digraph $G(A)$, defined in the following way. The digraph $G(A)$ has $m + n$ vertices and these are colored either black or white. There are m *black vertices*, in one-to-one correspondence with the rows of A, and they are denoted by the numbers $1, 2, \ldots, m$. There are n *white vertices*, in one-to-one correspondence with the columns of A, and they are denoted by $1, 2, \ldots, n$. There is an edge from each black vertex to each of the white vertices. Drawing the black vertices in a column and the white vertices in another column to the right, all edges are directed from left to right. To the edge going out from the black vertex i and terminating at the white vertex j we let correspond the matrix entry a_{ij}, where a_{ij} is called the *weight* of the edge. The digraph $G(A)$ is called the *König digraph* of the matrix A. The edges of the König digraph are in one-to-one correspondence with the positions of the matrix, with each edge *weighted* (or *labeled*) by the entry of A in the corresponding position. In summary, the vertices of a König digraph are of either color black or white, and the sets of black vertices and of white vertices have labels that are consecutive ordinal numbers beginning with 1; the edges can have any numbers as labels. Any digraph with these properties is the König digraph of a matrix. In fact, as should be clear, the König digraph is just an alternative structure to a rectangular array for viewing a matrix. The *type* of a König digraph is m by n (or $m \times n$), if there are m black vertices and n white vertices.

Example 2.2.1 The König digraph of the matrix

$$A = \begin{bmatrix} 1 & 2 & 3 \\ 4 & 5 & 6 \end{bmatrix}$$

is displayed in Figure 2.1. □

The digraph $G(A)$ of the matrix A is called the König digraph, because König used the corresponding bipartite graph in his papers (see [56]). In fact, D. König, the Hungarian mathematician who is considered to be the founder of modern graph theory, was the first to use graphical methods in matrix theory [55], [56]. There are many mathematical papers in which results from the matrix theory are obtained or proved by graph-theoretical means (see the Coda).

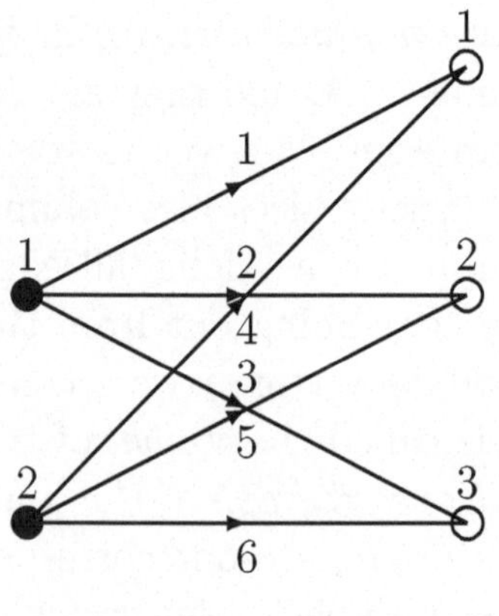

Figure 2.1

The matrix operations defined in the last section have counterparts for the König digraph, and we define these now.

Definition 2.2.2 Let G_1 and G_2 be two König digraphs.

1. *Digraph Sum*: Assume that G_1 and G_2 are of the same type. Then their sum $G_1 + G_2$ is the König digraph of that same type, where the weight of the edge from black vertex i to white vertex j is the sum of the weights of the corresponding edges of G_1 and G_2.

2. *Digraph Composition*: Let G_1 be of type m by n, and let G_2 be of type n by p. Then the number of white vertices of G_1 equals the number of black vertices of G_2. The composition $G_1 * G_2$ is the digraph of type m by p obtained by identifying each white vertex of G_1 with the correspondingly labeled black vertex of G_2. The digraph $G_1 * G_2$ has vertices of three colors: black (the black vertices of G_1), white (the

white vertices of G_2), and gray (the vertices obtained by identifying the white vertices of G_1 with the black vertices of G_2). Note that $G_1 * G_2$, having vertices of three different colors, is not a König digraph.

3. *Digraph Product:* Let G_1 be of type m by n, let G_2 be of type n by p, and consider the digraph composition $G_1 * G_2$. The product $G_1 \cdot G_2$ is the König digraph of type m by p whose black vertices are the black vertices of $G_1 * G_2$ and whose white vertices are the white vertices of $G_1 * G_2$. The weight of the edge from the ith black vertex to jth white vertex of $G_1 \cdot G_2$ equals the sum of the weights of all paths of length 2 between the ith black vertex and the jth white vertex of $G_1 * G_2$. (There are n such paths, and, in general, the *weight of a path* is the product of the weights of each of its edges.)

4. *Scalar Multiplication of a Digraph:* Let c be a scalar. Then $c \cdot G_1$ (or, sometimes, more simply, cG_1) is the digraph obtained from G_1 by multiplying the weight of each of its edges by c. □

Example 2.2.3 Two digraphs, G_1 and G_2, together with their composition $G_1 * G_2$ and product $G_1 \cdot G_2$, are displayed in Figure 2.2. In that figure, only the weights of some of the edges are given; namely, the weights of the edges leaving black vertex 1 in G_1, the weights of the edges terminating in white vertex 1 in G_2, and the weight of the edge from black vertex 1 to white vertex 1 in $G_1 \cdot G_2$. □

It should be clear that digraph addition and scalar multiplication of a digraph correspond to matrix addition and scalar multiplication of a matrix, respectively. More precisely, if A and B are matrices of the same size, then

$$G(cA) = cG(A) \text{ and } G(A+B) = G(A) + G(B).$$

An analogous conclusion holds for product, and we state and prove this in the next theorem.

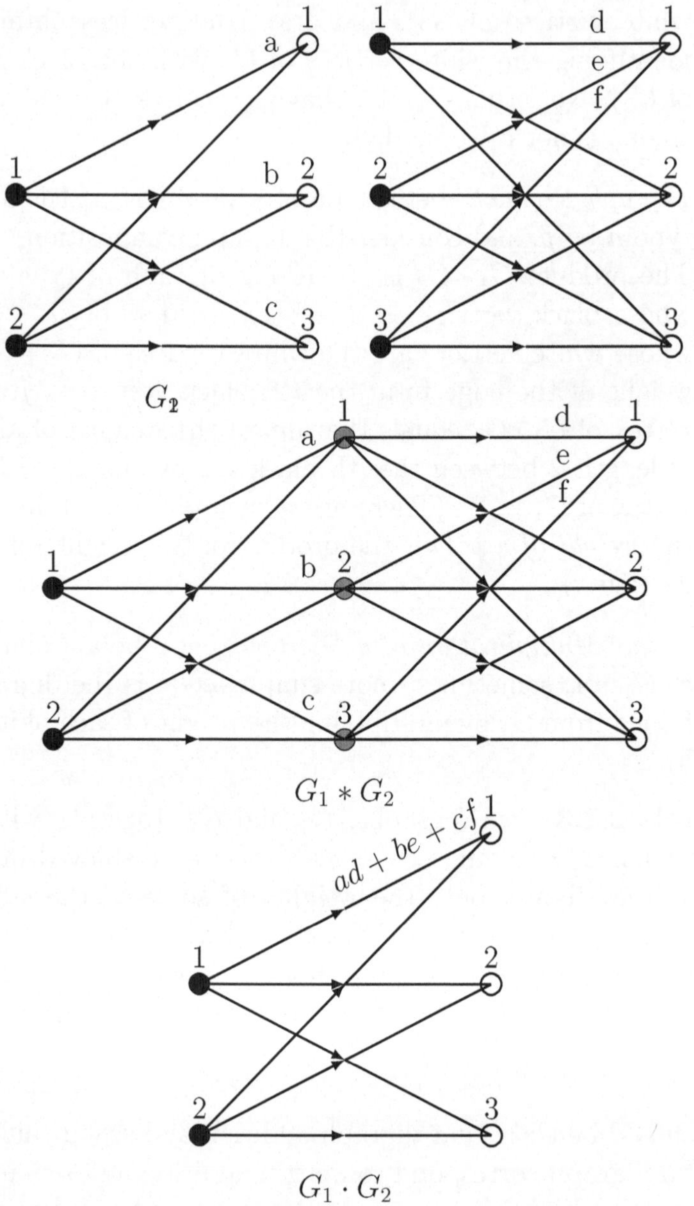

Figure 2.2

Theorem 2.2.4 *Let $A = [a_{ij}]$ be a matrix of type m by n, and let $B = [b_{ij}]$ be a matrix of type n by p. Then*

$$G(A \cdot B) = G(A) \cdot G(B).$$

Proof. In the composition $G(A) * G(B)$, there is a path of weight $a_{ik}b_{kj}$ from the ith black vertex to the jth white vertex that passes through the kth gray vertex for each $k = 1, 2, \ldots, n$. Hence the sum of the weights of all the paths of length 2 from the ith black vertex to the jth white vertex is

$$\sum_{j=1}^{n} a_{ij}b_{jk},$$

and this equals, according to the definition of a matrix product, the (i, j)-entry of AB. □

In the next theorem we collect some basic properties expressed in terms of graph operations and the corresponding matrix operations. In the theorem we assume that the types of the graphs and matrices are such that the operations can be carried out.

Theorem 2.2.5 *The following properties hold:*

1. *The composition of König digraphs is an associative operation:*
$$G_1 * (G_2 * G_3) = (G_1 * G_2) * G_3.$$
2. *The product of König digraphs is an associative operation:*
$$G_1 \cdot (G_2 \cdot G_3) = (G_1 \cdot G_2) \cdot G_3.$$
Equivalently, for matrices we have $A_1(A_2A_3) = (A_1A_2)A_3$.
3. *Graph multiplication is distributive over addition:*
$$\begin{aligned} G_1 \cdot (G_2 + G_3) &= G_1 \cdot G_2 + G_1 \cdot G_3 \qquad \textit{and} \\ (G_1 + G_2) \cdot G_3 &= G_1 \cdot G_3 + G_2 \cdot G_3. \end{aligned}$$
Equivalently, for matrices we have $A_1(A_2 + A_3) = A_1A_2 + A_1A_3$ *and* $(A_1 + A_2)A_3 = A_1A_3 + A_2A_3$.

Proof. These relations are readily verified. The equivalence of the distributive properties of graph multiplication and matrix multiplication are a consequence of

$$G(A) \cdot (G(B) + G(C)) = G(A) \cdot G(B) + G(A) \cdot G(C)$$

and

$$(G(A) + G(B)) \cdot G(C) = G(A) \cdot G(C) + G(B) \cdot G(C).$$

□

We now relate the König digraph to transposition. The König digraph $G(A^T)$ of the matrix A^T is obtained from the König digraph $G(A)$ of A by changing the color of black vertices to white, changing the color of white vertices to black, and then changing the orientation of all edges so that once again edges go from a black vertex to a white vertex.

Example 2.2.6 For the matrix

$$A = \begin{bmatrix} 1 & 2 & 3 \\ 4 & 5 & 6 \end{bmatrix},$$

given along with its König digraph in Example 2.2.1, the transpose A^T equals

$$A^T = \begin{bmatrix} 1 & 4 \\ 2 & 5 \\ 3 & 6 \end{bmatrix},$$

and its digraph is given in Figure 2.3. □

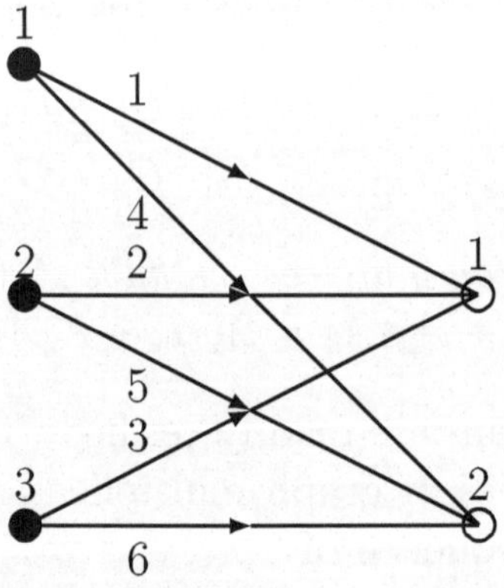

Figure 2.3

The anticommutativity property $(AB)^T = B^T A^T$ can be understood in terms of the König digraph. Consider the composition digraphs $G(A) * G(B)$ and $G(B^T) * G(A^T)$. If in $G(A) * G(B)$ we make the black vertices white and make the white vertices black and change the orientation of all edges, then we get the digraph $G(B^T) * G(A^T)$. This implies that $(AB)^T = B^T A^T$. Finally we note that using induction we get the more general product-transposition rule:

$$(A_1 A_2 \cdots A_k)^T = A_k^T \cdots A_2^T A_1^T \quad (k \geq 2).$$

To conclude this section we establish a convention that at times is helpful in both presentation and understanding. By definition, in a König digraph G there is an edge from each black vertex to each white vertex. However, if the weight of an edge is zero (corresponding to a zero entry in a matrix), then we can can just delete the edge from G. Thus, with this convention, *a König digraph has black vertices* $1, 2, \ldots, m$ *and white vertices* $1, 2, \ldots, n$ *and there are edges from some of the black vertices to some of the white vertices.* This convention does not influence our matrix calculations. For example, in the proof of Theorem 2.2.4, the path of length 2 in the composition $G(A) * G(B)$ from the ith black vertex to the jth white vertex that passes through the kth gray vertex disappears if either

(i) in $G(A)$ there is no edge from the ith black vertex to the kth white vertex (because $a_{ik} = 0$), or

(ii) in $G(B)$ there is no edge from the k black vertex to the jth white vertex (because $b_{kj} = 0$).

If (i) or (ii) holds, then the path of length 2 in the composition $G(A) * G(B)$ has weight $a_{ik} b_{kj} = 0$, and our calculation is not affected.

This same convention can be applied to any digraph with weighted edges.

Example 2.2.7 Consider the permutation matrix

$$P = \begin{bmatrix} 0 & 1 & 0 & 0 \\ 0 & 0 & 0 & 1 \\ 1 & 0 & 0 & 0 \\ 0 & 0 & 1 & 0 \end{bmatrix}$$

corresponding to the permutation $\sigma = 2413$ of $\{1, 2, 3, 4\}$. If we apply our convention, then the König digraph $G(P)$ has only four edges, each of weight equal to 1: an edge from black vertex 1 to white vertex 2, an edge from black vertex 2 to white vertex 4, an edge from black vertex 3 to white vertex 1, and an edge from black vertex 4 to white vertex 3. In general, the König digraph of a permutation matrix of order n corresponding to the permutation $\sigma = k_1 k_2 \ldots k_n$ has n edges of weight 1, namely, the edges from black vertex i to white vertex k_i $(i = 1, 2, \ldots, n)$. There is exactly one edge beginning at each black vertex and exactly one edge terminating at each white vertex; these edges can be regarded as defining a one-to-one corespondence between the black vertices and the white vertices. □

Using our convention illuminates the proof of the following basic fact.

Theorem 2.2.8 *The product of two permutation matrices of the same order n is also a permutation matrix of order n.*

Proof. Let P and Q be the permutation matrices corresponding to the permutations $\sigma = k_1 k_2 \cdots k_n$ and $\pi = l_1 l_2 \cdots l_n$, respectively. Then, in $G(P * Q)$, there are exactly n paths of length 2 from black vertices to white vertices, each of weight equal to $1 \cdot 1 = 1$, and these paths have no vertices in common. In $G(PQ)$, there is exactly one edge from each black vertex to each white vertex, and these edges all have weight equal to 1. More precisely, for each $i = 1, 2, \ldots, n$, there is an edge of weight 1 from black vertex i to white vertex l_{k_i}. Because $k_1 k_2 \cdots k_n$ and $l_1 l_2 \cdots l_n$ are both permutations of $\{1, 2, \ldots, n\}$, $l_{k_1} l_{k_2} \cdots l_{k_n}$ is also a permutation of $\{1, 2 \ldots, n\}$. Thus PQ is a permutation matrix. □

2.3 Partitioned Matrices

A matrix A may be partitioned into smaller matrices by inserting horizontal and vertical lines that partition its set of rows and its set of columns, respectively. Such a matrix is then called a *partitioned* or *block matrix*, with the resulting smaller matrices called *blocks*.

Example 2.3.1 Let

$$A = \begin{bmatrix} 1 & 2 & 3 & 4 \\ 5 & 6 & 7 & 8 \\ 8 & 7 & 6 & 5 \\ 4 & 3 & 2 & 1 \end{bmatrix}.$$

We can, for instance, partition A into blocks in the following way:

$$A = \left[\begin{array}{c|cc|c} 1 & 2 & 3 & 4 \\ \hline 5 & 6 & 7 & 8 \\ 8 & 7 & 6 & 5 \\ 4 & 3 & 2 & 1 \end{array}\right].$$

The blocks of A are the matrices

$$B_1 = [1], \quad B_2 = [2\ 3], \quad B_3 = [4]$$

$$B_4 = \begin{bmatrix} 5 \\ 8 \\ 4 \end{bmatrix}, \quad B_5 = \begin{bmatrix} 6 & 7 \\ 7 & 6 \\ 3 & 2 \end{bmatrix}, \quad B_6 = \begin{bmatrix} 8 \\ 5 \\ 1 \end{bmatrix}.$$

Using these blocks we may write A as

$$A = \begin{bmatrix} B_1 & B_2 & B_3 \\ B_4 & B_5 & B_6 \end{bmatrix}.$$

□

Example 2.3.2 The matrix

$$A = \begin{bmatrix} O & I_2 \\ I_2 & O \end{bmatrix}$$

is, in fact, the permutation matrix

$$A = \begin{bmatrix} 0 & 0 & 1 & 0 \\ 0 & 0 & 0 & 1 \\ 1 & 0 & 0 & 0 \\ 0 & 1 & 0 & 0 \end{bmatrix}.$$

□

If a matrix is partitioned into blocks by partitioning its rows into μ nonempty sets and its columns into ν nonempty sets, then it is of the form

$$A = \begin{bmatrix} A_{11} & A_{12} & \cdots & A_{1\nu} \\ A_{21} & A_{22} & \cdots & A_{2\nu} \\ \vdots & \vdots & \ddots & \vdots \\ A_{\mu 1} & A_{\mu 2} & \cdots & A_{\mu\nu} \end{bmatrix}, \tag{2.4}$$

where the A_{ij} are the blocks of the matrix partition. The partitioned matrix A in (2.4) can be regarded as a matrix of type μ by ν whose entries are themselves matrices (the blocks). We say that A is a *block matrix of type* μ *by* ν. Using these ideas we can carry over our basic matrix operations to partitioned matrices.

If c is a number, then evidently

$$cA = \begin{bmatrix} cA_{11} & cA_{12} & \cdots & cA_{1\nu} \\ cA_{21} & cA_{22} & \cdots & cA_{2\nu} \\ \vdots & \vdots & \ddots & \vdots \\ cA_{\mu 1} & cA_{\mu 2} & \cdots & cA_{\mu\nu} \end{bmatrix}.$$

If B is a matrix of the same type as A, and B is partitioned into blocks in the same way that A is partitioned in (2.4), then the definition of matrix addition implies that

$$A + B = \begin{bmatrix} A_{11} + B_{11} & A_{12} + B_{12} & \cdots & A_{1\nu} + B_{1\nu} \\ A_{21} + B_{21} & A_{22} + B_{22} & \cdots & A_{2\nu} + B_{2\nu} \\ \vdots & \vdots & \ddots & \vdots \\ A_{\mu 1} + B_{\mu 1} & A_{\mu 2} + B_{\mu 2} & \cdots & A_{\mu\nu} + B_{\mu\nu} \end{bmatrix}.$$

The relationship of matrix multiplication to partitioned matrices is a little more subtle.

Now let $A = [a_{ij}]$ and $B = [b_{ij}]$ be matrices of types m by n and n by p, respectively. Assume that A and B are partitioned as block matrices of types μ by ν and ν by λ, respectively:

$$A = \begin{bmatrix} M_{11} & M_{12} & \cdots & M_{1\nu} \\ M_{21} & M_{22} & \cdots & M_{2\nu} \\ \vdots & \vdots & \ddots & \vdots \\ M_{\mu 1} & M_{\mu 2} & \cdots & M_{\mu\nu} \end{bmatrix}, \quad B = \begin{bmatrix} N_{11} & N_{12} & \cdots & N_{1\lambda} \\ N_{21} & N_{22} & \cdots & N_{2\lambda} \\ \vdots & \vdots & \ddots & \vdots \\ N_{\nu 1} & N_{\nu 2} & \cdots & N_{\nu\lambda} \end{bmatrix}.$$

Assume also that the column partition of A agrees with the row partition of B. This means that M_{ik} is an m_i by n_k matrix and N_{kj} is an n_k by p_j matrix. Here the integers m, n, and p are *partitioned* as $m = m_1 + m_2 + \cdots + m_\mu, n = n_1 + n_2 + \cdots + n_\nu$, and $p = p_1 + p_2 + \cdots + p_\lambda$. Under these circumstances, we say that A and B are *conformally partitioned.*

Let the set of black vertices of $G(A)$ be partitioned in accordance with the partition of the integer m, and let the set of white vertices of $G(A)$ be partitioned according to the partition of the integer n. Similarly, let the black and white vertices of $G(B)$ be partitioned according to partitions for n and p, respectively. In forming $G(A) * G(B)$ and $G(A) \cdot G(B)$, one gets a natural partitioning for the product AB as

$$AB = \begin{bmatrix} P_{11} & P_{12} & \cdots & P_{1\lambda} \\ P_{21} & P_{22} & \cdots & P_{2\lambda} \\ \vdots & \vdots & \ddots & \vdots \\ P_{\mu 1} & P_{\mu 2} & \cdots & P_{\mu\lambda} \end{bmatrix}$$

where the blocks P_{ij} are of size m_i by p_j.

Now we have to see how to calculate the block P_{ij}. The paths of length 2 that in $G(A) * G(B)$ start from the set of m_i black vertices and terminate in the set of p_j white vertices correspond to block P_{ij} of AB. These paths, of course, cross through gray vertices. We partition the n gray vertices into ν parts according to the partition $n = n_1 + n_2 + \cdots + n_\nu$ of n. The paths of length 2 that cross through the set of n_k gray vertices correspond to the matrix product $M_{ik}N_{kj}$. Because each path of length 2 in $G(A) * G(B)$ crosses through exactly one of the ν sets of gray vertices, we obtain the formula

$$P_{ij} = M_{i1}N_{1k} + M_{i2}N_{2k} + \cdots + M_{i\nu}N_{\nu k}. \tag{2.5}$$

In other words, matrices conformally partitioned into blocks are multiplied formally by the same rule as for matrix multiplication.

2.4 Exercises

1. Compute the matrix product

$$\begin{bmatrix} 1 & 3 \\ 2 & 5 \\ 4 & 0 \end{bmatrix} \begin{bmatrix} 2 & 3 & 4 \\ 1 & 5 & 1 \end{bmatrix}.$$

2. Let $D = \mathrm{diag}(d_1, d_2, \ldots, d_n)$, and let A be a matrix of order n. Show that DA is the matrix obtained from A by multiplying each element in row i by d_i for $i = 1, 2, \ldots, n$, and that for AD we multiply each element in column i by d_i.

3. Let A be an m by n matrix and let B be an n by p matrix. Let $\alpha_1, \alpha_2, \ldots, \alpha_m$ be the rows of A and let $\gamma_1, \gamma_2, \ldots, \gamma_p$ be the columns of B. Show that the rows of AB are

$$\alpha_1 B, \alpha_2 B, \ldots, \alpha_m B$$

and the columns of AB are

$$A\gamma_1, A\gamma_2, \ldots, A\gamma_p.$$

Conclude that if A has a row of all zeros, so does AB, and that if B has a column of all zeros, so does AB.

4. Let A and B be upper (lower) triangular matrices of order n. By using the König digraph, show that the matrix AB is also upper (lower) triangular matrix.

5. Construct the König digraphs of the two matrices in Exercise 1, and compute their digraph composition and product.

6. Let

$$A = \begin{bmatrix} a_{11} & a_{12} & a_{13} & a_{14} & a_{15} \\ a_{21} & a_{22} & a_{23} & a_{24} & a_{25} \\ a_{31} & a_{32} & a_{33} & a_{34} & a_{35} \\ a_{41} & a_{42} & a_{43} & a_{44} & a_{45} \end{bmatrix}.$$

Let P be the permutation matrix corresponding to the permutation 2341 of $\{1, 2, 3, 4\}$, and let Q be the permutation matrix corresponding to the permutation 43512 of $\{1, 2, 3, 4, 5\}$. Using the König digraph, compute PAQ.

7. Let r be a nonnegative integer, and define $H_r = [h_{ij}^{(r)}]$ to be the matrix of order n with $h_{ij}^{(r)} = \delta_{i,j-r+1}$ $(1 \leq i, j \leq n)$. (Here the subscript $j - r + 1$ is whichever of $1, 2, \ldots, n$ it is congruent to modulo n.) First show that H_r is a permutation matrix, and then, using the König digraph, show that $H_p H_q = H_{p+q}$ whenever p and q are nonnegative integers.

8. Let $I_n(i, j)$ be the (permutation) matrix of order n obtained by interchanging rows i and j of the identity matrix I_n. Thus, $I_n(i, j) = I_n(j, i)$. Show that the following identities hold:

$$I_n(i, j)^2 = I_n \text{ and } I_n(i, k)I_n(k, j)I_n(j, i) = I_n(k, j).$$

9. Let P be a permutation matrix of order n. Use the König digraph to prove that

$$PP^T = P^T P = I_n.$$

10. Using block multiplication, compute the product

$$\left[\begin{array}{c|c|c} I_2 & O_2 & I_2 \\ \hline O_2 & I_2 & I_2 \end{array}\right] \left[\begin{array}{c|c} I_2 & I_2 \\ \hline O_2 & -I_2 \\ \hline -I_2 & I_2 \end{array}\right].$$

Chapter 3

Powers of Matrices

In this chapter we consider powers of square matrices and describe them in terms of a digraph, different from the König digraph, that we associate with a square matrix. The basic result here is a theorem by which the entries of powers of a square matrix can be calculated by enumeration of certain walks in the associated digraph. As applications we consider Markov chains, finite automata, and counting permutations with certain restrictions. We also show how a certain structured matrix called a circulant results from the powers of a matrix whose digraph is a cycle.

3.1 Matrix Powers and Digraphs

An *associative groupoid* is a pair $(X, \cdot)$ consisting of a nonempty set X and a binary operation, denoted by the usual multiplication symbol $\cdot$, that satisfies the associative law. The associative groupoid may have an *identity element* e satisfying $a \cdot e = e \cdot a = a$ for every element a in X. In an associative groupoid $(X, \cdot)$, for every element $a \in X$ and every nonnegative integer k, the *kth power* of a is defined inductively as follows:

$$a^k = \begin{cases} a, & \text{if } k = 1, \\ a \cdot a^{k-1}, & \text{if } k > 1. \end{cases}$$

If $(X, \cdot)$ has an identity element e, then we also set $a^0 = e$.

The set of square matrices of a given order n over a field is an associative groupoid under multiplication with identity element equal to the identity matrix I_n. Thus nonnegative integral powers of a matrix A of order n are defined by $A^0 = I_n$, $A^1 = A$, and $A^k = A \cdot A^{k-1}$ for $k > 1$.

Now that we have defined matrix powers, we can also define matrix polynomials in a natural way. If

$$p(x) = a_0 x^k + a_1 x^{k-1} + \cdots + a_{k-1}x + a_k$$

is a polynomial of degree k, where the coefficients $a_0, a_1, \ldots, a_k$ are elements of a field F and A is a square matrix over F, then we define the *matrix polynomial* $p(A)$ (really, the evaluation of a polynomial at a square matrix) to be

$$p(A) = a_0 A^k + a_1 A^{k-1} + \cdots + a_{k-1}A + a_k I.$$

Example 3.1.1 Let

$$A = \begin{bmatrix} 1 & 2 \\ 3 & 5 \end{bmatrix} \text{ and } p(x) = 2x^2 - x + 3.$$

Then

$$\begin{aligned} p(A) = 2A^2 - A + 3I_2 &= 2\begin{bmatrix} 7 & 12 \\ 18 & 31 \end{bmatrix} - \begin{bmatrix} 1 & 2 \\ 3 & 5 \end{bmatrix} + \begin{bmatrix} 1 & 0 \\ 0 & 1 \end{bmatrix} \\ &= \begin{bmatrix} 7 & 10 \\ 15 & 27 \end{bmatrix}. \end{aligned}$$

□

Let $A = [a_{ij}]$ be a matrix of order n. We associate with A a *digraph* $D(A)$ with n vertices. The vertices of $D(A)$ are denoted by $1, 2, \ldots, n$. (Unlike the König digraph, the vertices correspond simultaneously to the n rows and the n columns of A.) There is an edge from vertex i to vertex j of weight a_{ij} for each $i, j = 1, 2, \ldots, n$. Thus $D(A)$ has a loop at each vertex i of weight a_{ii}. As with the König digraph, *an edge of weight zero, corresponding to a zero entry of A, can be removed from $D(A)$ without any effect* *in our subsequent calculations*; indeed, removing such edges can have the effect of making certain calculations more transparent as it reveals more clearly the structure of the digraph. The *weight of a walk* in $D(A)$ is defined to be the product of the weights of all edges of the walk.[1] Powers of a matrix can be calculated using the digraph $D(A)$.

Theorem 3.1.2 *Let $A = [a_{ij}]$ be a matrix of order n. For each positive integer k, the entry $a_{ij}^{(k)}$ of A^k in the ith row and jth column equals the sum of the weights of all walks in $D(A)$ of length k from vertex i to vertex j.*

Proof. We shall give two proofs of this result: the first uses directly the digraph $D(A)$, and the second uses the König digraph $G(A)$ in an auxiliarly way.

First proof: We proceed by induction on k. If $k = 1$, the theorem is a direct consequence of the definition of the digraph $D(A)$. This is because for each i and j, there is exactly one walk of length 1 from i to j and it has weight a_{ij}. Now assume the theorem holds for the integer k. By the definition of matrix powers, $A^{k+1} = A \cdot A^k$, and so by matrix multiplication we get

$$a_{ij}^{(k+1)} = a_{i1}a_{1j}^{(k)} + a_{i2}a_{2j}^{(k)} + \cdots + a_{in}a_{nj}^{(k)} = \sum_{r=1}^{n} a_{ir}a_{rj}^{(k)}.$$

By the inductive assumption, for each $r = 1, 2, \ldots, n$, $a_{rj}^{(k)}$ is the sum of the weights of all walks of length k in $D(A)$ from vertex r to vertex j. Consider a walk γ of length $k+1$ from vertex i to vertex j. The walk γ consists of an edge from i to r for some r between 1 and n, followed by a walk γ' of length k from r to j. The weight of γ equals a_{ir} times the weight of γ'. Conversely, a walk γ' of length k from r to j preceded by the edge from i to r gives a walk γ of length $k+1$ from i to j whose weights satisfy this

[1]Now we see the advantage of suppressing edges of weight zero. The weight of a walk that contains an edge of weight zero is zero. If we suppress the edge of weight zero, then the walk "disappears" and so makes no contribution to a sum.

same rule. It follows that $a_{ir}a_{rj}^{(k)}$ equals the weight of all walks of length k from i to j whose first edge is an edge from i to r. Hence $\sum_{r=1}^{n} a_{ir}a_{rj}^{(k)}$ is the sum of the weights of all walks of length $k+1$ from i to j, completing the induction and proving the theorem.

Second proof: Here we will be more brief. Consider the digraph $G(A)^{(k)} = G(A) * G(A) * \cdots * G(A)$ equal to the composition of k copies of the König digraph $G(A)$. This digraph has $k+1$ sets of n vertices with the first set black, the last set white, and all others gray. The sum of the weights of all walks from black vertex i to white vertex j equals $a_{ij}^{(k)}$. There is a one-to-one correspondence between walks of length k in $G(A)^{(k)}$ from black vertex i to white vertex j and walks of length k in $D(A)$ from its vertex i to its vertex j. Moreover, corresponding walks have the same weight. Thus $a_{ij}^{(k)}$ equals the sum of the weights of all walks of length k in $D(A)$ from i to j. □

We give an example that demonstrates the usefulness of Theorem 3.1.2. Omitting the edges of weight zero (if there are, relatively speaking, many such edges) allows one to identify walks in a digraph more readily.

Example 3.1.3 The digraph $D(A)$ corresponding to the matrix

$$A = \begin{bmatrix} a & b \\ 0 & c \end{bmatrix}$$

is drawn in Figure 3.1.

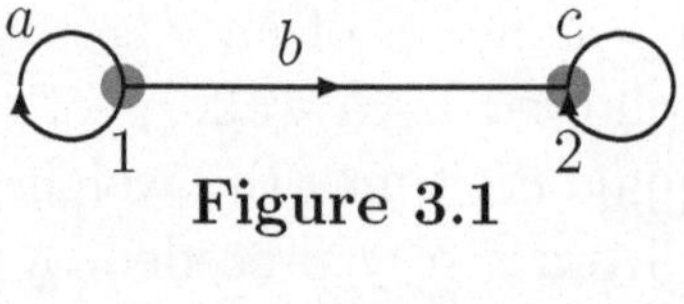

Figure 3.1

From the vertex 1 to the vertex 1 there is only one walk of length k, and its weight is a^k. Similarly, there is only one walk of length k from vertex 2 to itself and it has weight c^k. From 1 to 2 there are k walks of length k. These are the walks of weight $a^i bc^{k-1-i}$

consisting of i loops of weight a, followed by the edge of weight b, and then $k-1-i$ loops of weight c. Here i can be any integer from 0 to $k-1$. Therefore, the element on the position $(1,2)$ of the matrix A^k is equal to $\sum_{i=0}^{k-1} a^i c^{k-1-i}$. From the vertex 2 to the vertex 1 there are no walks. Hence we have

$$A^k = \begin{bmatrix} a^k & b(a^{k-1} + a^{k-2}c + \cdots + c^{k-1}) \\ 0 & c^k \end{bmatrix}.$$

□

A square matrix is *nilpotent* provided there is a positive integer k such that $A^k = 0$. Note that it follows from the inductive definition of matrix powers that if $A^k = O$, then $A^r = O$ for all $r \geq k$. An arbitrary matrix is *nonnegative* if all its entries are nonnegative numbers. Using Theorem 3.1.2, we can characterize nonnegative matrices A that are nilpotent.

Theorem 3.1.4 *Let A be a square matrix of order n. Then A is nilpotent if the corresponding digraph $D(A)$ does not have any cycles; in this case, $A^n = O$. A nonnegative square matrix A is nilpotent if and only if the corresponding digraph $D(A)$ does not have any cycles.*

Proof. Applying Theorem 3.1.2, we see that A is a nilpotent matrix if and only if there exists a positive integer k such that the digraph $D(A)$ contains no walk of nonzero weight of length r for all $r \geq k$. If $D(A)$ does not have a cycle, then there can be no such walk of length n or greater, since such a walk would repeat a vertex and thus create a cycle. Hence A is nilpotent and $A^n = O$ if $D(A)$ does not have any cycles.

Now suppose that A is a nonnegative matrix and A is nilpotent. If $D(A)$ contains a cycle, then $D(A)$ has walks of arbitrary long length of positive weight, contradicting the assumption that A is nilpotent. □

Note that if a matrix A of order n is nilpotent, then $A^r = O$ for all $r \geq n+1$. This is because in the digraph $D(A)$ with n vertices, if there is a closed walk, then there is a cycle of length at most n.

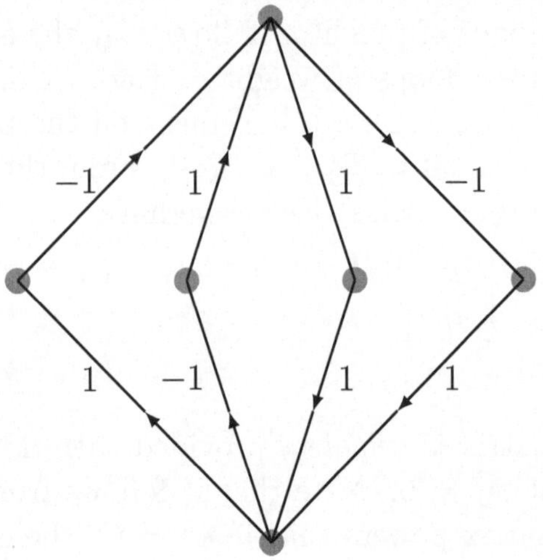

Figure 3.2

Example 3.1.5 A *strictly upper triangular matrix* is an upper triangular matrix that also has only zeros on its main diagonal. Thus a matrix $A = [a_{ij}]$ of order n is a strictly upper triangular matrix if and only if $a_{ij} = 0$ for all i and i with $j \leq i$. Let A be a strictly upper triangular matrix. Then its digraph $D(A)$ does not have any cycles since all edges go from a vertex i to a vertex j with $j > i$. By Theorem 3.1.4 the matrix A is nilpotent. □

Example 3.1.6 Theorem 3.1.4 asserts, in particular, that the digraph of a nonnegative nilpotent matrix cannot have a cycle. The assumption that the matrix is nonnegative cannot be omitted. The digraph of the matrix

$$\left[\begin{array}{rrrrrr} 0 & 0 & 0 & 0 & 1 & -1 \\ 0 & 0 & 1 & -1 & 0 & 0 \\ -1 & 0 & 0 & 0 & 0 & 0 \\ 1 & 0 & 0 & 0 & 0 & 0 \\ 0 & 1 & 0 & 0 & 0 & 0 \\ 0 & 1 & 0 & 0 & 0 & 0 \end{array}\right]$$

is displayed in Figure 3.2. The matrix is nilpotent, yet the digraph has cycles.[2] Of course, if $D(A)$ does not have any cycles, then A

[2]This example was constructed by Z. Lukić.

is nilpotent, but, as the example shows, the converse is not true in general. □

Example 3.1.7 Let a digraph be obtained from a tree by orienting its edges in an arbitrary way, and assign weights to the edges in any way whatsoever. Consider a matrx A with $D(A)$ equal to this digraph. Because $D(A)$ does not contain cycles, the matrix must be nilpotent. For example, the matrices

$$\begin{bmatrix} 0 & -2 & 0 & 0 \\ 0 & 0 & 1 & 0 \\ 0 & 0 & 0 & 5 \\ 0 & 0 & 0 & 0 \end{bmatrix} \text{ and } \begin{bmatrix} 0 & 1 & 0 & 0 \\ 0 & 0 & 0 & 0 \\ 0 & 3 & 0 & 0 \\ 0 & 0 & 2 & 0 \end{bmatrix}$$

are constructed from a path of 4 vertices in this way and are nilpotent. □

It follows from Theorem 3.1.4 that whether or not a *nonnegative* matrix is nilpotent depends only on the structure of the digraph and not on the weights of its edges. There are also other properties of matrices that depend only on which elements are 0 and do not depend on the values of the elements different from 0. Such characteristics are described in a natural way by means of digraphs.

We next describe an application of Theorem 3.1.2 in probability theory.

Example 3.1.8 Some random processes can be described by the following model:

Let G be a digraph with n vertices, containing all possible edges, including a loop at each vertex. Consider the vertices to be *states* of a system, and imagine that an object, let us call it a *particle*, moves in a random way along the edges of the digraph in the direction of the edge. Normally, the particle is on a vertex of the digraph and at moments of time $t = 1, 2, \ldots$ moves along an edge to another vertex or the same vertex in case of a loop. If the particle at some initial moment $t = t_0$ is on a vertex i (in state i), at the next moment $t = t_0 + 1$ it has moved to vertex j (to state

j) with probability $p_{ij} \geq 0$, $(i, j = 1, 2, \ldots, n)$. The values p_{ij} are independent of the value of the discrete time variable t and, in order that we have probability distributions for the transition at each vertex, the p_{ij} satisfy the condition

$$p_{i1} + p_{i2} + \cdots + p_{in} = 1 \quad (i = 1, 2, \ldots, n).$$

In particular, the digraph of P satisfies $D(P) = G$.[3]

The matrix $P = [p_{ij}]$ of order n is called the *one-step transition probability matrix*. We call such a digraph G a *Markov chain*, and indeed use this term for the matrix P itself.

It is interesting and important to investigate the behavior of a Markov chain over a long period of time. Of special interest are those cases when some of the values p_{ij} are equal to 0. When drawing the digraph G, we may omit an edge from a vertex i to a vertex j of weight 0, since the particle cannot move from i to j in this case. In this way, and as we have discussed earlier, the structure of the digraph, and hence the important characteristics of the Markov chain, become clearer in the reduced digraph.

Because the values of the transition matrix P are the same for all times, the probability that the particle gets from a vertex i in k steps to a vertex j along a fixed walk (of length k) is equal to the product of weights of the edges along that walk, i.e. to the weight of the walk. According to Theorem 3.1.2, we conclude that the probability of getting from a state i to a state j in k steps (along any walk) equals the element $p_{ij}^{(k)}$ in position (i, j) of the matrix P^k. Therefore, the behavior of a Markov chain is determined by the structure of the matrices P^k $(k = 1, 2, \ldots)$. Almost all interesting characteristics of a matrix P^k can be determined by means of the structure of the corresponding digraph, while the weights of the edges affect only the quantitative characteristics of the Markov chain (see Chapter 8). □

[3]Here is an amusing formulation of this problem. Think of the digraph as a map of a city and the particle as a drunkard who is trying to get home (one of the vertices of the digraph). At each intersection, he chooses one of the streets to take according to the given probabilities. The question arises as to whether the drunkard will reach home (of course, depending on his level of inebriation, he may or may not recognize his home!). It turns out that under some mild conditions, the drunkard will reach home with high probability.

ection, we discuss briefly an application of automata theory.

automaton is a map from finite sequences symbols (the *input symbols*) into finite inite set of symbols (the *output symbols*). A omaton can be represented by a digraph G. The vertices 1,2,..., n of the digraph represent the states of the automaton in discrete moments of time $t = 0, 1, 2, \ldots$. If in some moment of time the automaton is in a state i, and if it is affected by a symbol x_j of the input alphabet, the automaton goes to a new state determined by i and x_j, while a symbol of the output alphabet, also determined by the i and x_j, appears at the output. Let $X = \{x_1, x_2, \ldots, x_n\}$ be the set of all input symbols. In the following we consider the input symbols $x_1, x_2, \ldots, x_n$ as variables. We extend this set by the empty symbol with the meaning that at a given moment there is no symbol affecting the input of the automaton. This empty symbol is denoted by 0, and we will sometimes, according to need, interpret it as the number 0. If the symbols $x_{i_1}, x_{i_2}, \ldots, x_{i_s}$ are those that turn the automaton from a state i to a state j, the edge of the digraph joining the vertices i and j gets the following sum as its weight:

$$a_{ij} = x_{i_1} + x_{i_2} + \cdots + x_{i_s}. \tag{3.1}$$

If $a_{ij} = 0$, then this means that it is impossible to get from the state i to the state j, and the automaton stays in state i. In analogy with Markov chains, the matrix $A = [a_{ij}]$ of order n is called the *transition matrix of the automaton.*

Consider a walk of a length k between vertices i and j. The weight of that walk is the product of values of the form (3.1). If we multiply k sums of the type (3.1), we get a sum of products, every summand being the product of k members of the set X. We assume that the multiplication of the elements of X is noncommutative, so that the k factors in every product maintain their original order. If $x_{j_1} x_{j_2} \cdots x_{j_k}$ is some summand from the weight of the walk, the sequence of input symbols $x_{j_1}, x_{j_2}, \ldots, x_{j_k}$ sends the automaton from the state i to the state j. Therefore, the sum of the weights of all walks of length k between vertices i and j

produces the set of all sequences of k input symbols that automaton from the vertex i to the vertex j. According to orem 3.1.2, the element in position (i, j) of the matrix A^k is sum of terms, where each term is the product of k elements of the set X, and every such product determines a sequence of k symbols that sends the automaton from the state i to the state j. □

3.2 Circulant Matrices

A *circulant* is a square matrix of the form

$$A = \begin{bmatrix} a_0 & a_1 & a_2 & \cdots & a_{n-2} & a_{n-1} \\ a_{n-1} & a_0 & a_1 & \cdots & a_{n-3} & a_{n-2} \\ a_{n-2} & a_{n-1} & a_0 & \cdots & a_{n-4} & a_{n-3} \\ \vdots & \vdots & \vdots & \ddots & \vdots & \vdots \\ a_2 & a_3 & a_4 & \cdots & a_0 & a_1 \\ a_1 & a_2 & a_3 & \cdots & a_{n-1} & a_0 \end{bmatrix}. \tag{3.2}$$

Each row of such a matrix is a cyclic permutation of the first row; each column is a cyclic permuation of the first column.

There is a representation of a circulant (3.2) in terms of powers of a certain permutation matrix. Let P be the permutation matrix of order n defined by

$$P = \begin{bmatrix} 0 & 1 & 0 & \cdots & 0 & 0 \\ 0 & 0 & 1 & \cdots & 0 & 0 \\ 0 & 0 & 0 & \cdots & 0 & 0 \\ \vdots & \vdots & \vdots & \ddots & \vdots & \\ 0 & 0 & 0 & \cdots & 0 & 1 \\ 1 & 0 & 0 & \cdots & 0 & 0 \end{bmatrix} = \begin{bmatrix} O & I_{n-1} \\ I_1 & O \end{bmatrix}.$$

The digraph $D(P)$ corresponding to P is a cycle with n vertices and n edges from vertex i to vertex $i+1$ for $i = 1, 2, \ldots, n$, where $n+1$ is treated as 1 (i.e., computed modulo n, taking as residues $1, 2, \ldots, n$). For each positive integer k and each vertex i there is exactly one walk of length k beginning at i, and it terminates at vertex j mod n. Thus we have that

$$P^k = \begin{bmatrix} O & I_{n-k} \\ I_k & O \end{bmatrix} \quad (k = 1, 2, \ldots, n-1, n),$$

in particular, $P^n = I_n$. Hence we obtain a representation of the circulant given in (3.2) as

$$A = a_0 I + a_1 P + a_2 P^2 + \cdots + a_{n-1} P^{n-1}.$$

If we define the polynomial

$$g(x) = a_0 + a_1 x + a_2 x^2 + \cdots + a_{n-1} x^{n-1},$$

then

$$A = g(P).$$

Let $f(x)$ be any polynomial and divide $f(x)$ by $x^n - 1$ to get

$$f(x) = q(x)(x^n - 1) + r(x),$$

where the remainder $r(x)$ is a polynomial of degree at most $n-1$ (including possibly the zero polynomial). Since $P^n = I_n$, it follows that

$$f(P) = r(P),$$

implying that circulants are precisely the matrices that are polynomials in the permutation matrix P.

3.3 Permutations with Restrictions

Let $X = \{x_1, x_2, \ldots, x_n\}$ be a set of n elements. As we saw in Section 1.3, an *r-permutation-with-repetition* of X is an ordered arrangement of r elements of X with repetition of elements permitted, that is, an r-tuple $x_{i_1} x_{i_2} \ldots x_{i_k}$, where $1 \le i_j \le n$ for $j = 1, 2, \ldots, n$.

When we form permutations, we may impose certain restrictions. Here we consider restrictions of a very special type. Assume that for each $i = 1, 2, \ldots, n$, the set X is partitioned into two sets, X_i^1 and X_i^2. Thus

$$X = X_i^1 \cup X_i^2 \text{ where } X_i^1 \cap X_i^2 = \emptyset, \quad (i = 1, 2, \ldots, n).$$

We now require that the element x_i, wherever it occurs in the permutation, be followed, if it is not the last element in the permutation, by an element of X_i^1 only $(i = 1, 2, \ldots, n)$. Thus a pair x_i, x_j of adjacent elements in a permutation is a *permitted pair* provided $x_j \in X_i^1$. Define a matrix $A = [a_{ij}]$ of order n, where $a_{ij} = 1$ if x_i, x_j is a permitted pair, and $a_{ij} = 0$ otherwise. The matrix A is the *matrix of the permitted pairs.* The matrix $\overline{A}$ obtained from A by replacing 0's with 1's, and viceversa, is the *restriction matrix.*

We now determine the number $p_{n,k}(A)$ of k-permutations-with-repetition of X if a matrix A of permitted pairs is given. Since A is a matrix of 0's and 1's, all edges, and hence all walks, of the digraph $D(A)$ have weight 1. Thus the sum of the weights of the walks of length k from vertex i to vertex j in $D(A)$ equals the number of such walks. By Theorem 3.1.2, the number of walks of a length k from vertex i to vertex j equals the element $a_{ij}^{(k)}$ in the ith row and the jth column in the matrix A^k.

Denote the sum of all elements of a matrix Y by $\Sigma(Y)$. We thus have the formula

$$p_{n,k}(A) = \Sigma(A^{k-1}), \quad (k \geq 1).$$

3.4 Exercises

1. Let

$$A = \begin{bmatrix} 1 & 2 & 3 & 0 \\ 0 & 4 & 1 & 4 \\ 2 & 0 & 3 & 2 \\ 1 & 3 & 0 & 1 \end{bmatrix}.$$

Use the digraph $D(A)$ to compute A^2, A^3, and A^4.

2. Let $A = [a_{ij}]$ be the matrix of order n defined by $a_{ij} = \delta_{i,j-1}$ $(i, j = 1, 2, \ldots, n)$, and let $B = aI_n + A$, where a is

some constant. For k a positive integer, use the digraph of a matrix to compute A^k and B^k.

3. Let $A = [a_{ij}]$ be a matrix of order n such that all entries of A satisfy $|a_{ij}| \leq r$. Let k be a positive integer. By bounding the number of walks of length k in the digraph $D(A)$, show that the entries $a_{ij}^{(k)}$ of A^k satisfy $|a_{ij}^{(k)}| \leq n^{k-1}r^k$.

4. Let D be the digraph obtained from the complete graph K_5 (no loops) by replacing each edge with two oppositely directed edges. At each discrete time t, a particle always moves to a different vertex with equal probability $1/4$. Compute both the one-step and two-step transition probability matrices P and P^2.

5. Let D be a digraph obtained from K_5 by orienting each edge (your choice how to orient). At each discrete time t, a particle chooses one of the edges leaving its current location with equal probability. As in the previous exercise, compute both the one-step and two-step transition probability matrices P and P^2.

6. For each of the three trees of order 5 (see Section 1.1), give an orientation to each of the edges and construct (and verify) a nilpotent matrix A such that $D(A)$ is the resulting digraph.

7. Show that the product of two circulants of order n is a circulant.

8. Show that the transpose of a circulant is a circulant.

Chapter 4

Determinants

In this chapter we first define the Coates digraph of a square matrix. The Coates digraph is a slight variation of the digraph used in the previous chapter. We use the Coates digraph to give a nontraditional definition of the determinant of a square matrix. Using this definition, we derive the basic properties of a determinant that are useful in its evaluation. In particular, it is shown how the calculation of a determinant can be reduced to the calculation of determinants of lower order. We also derive the formula for the determinant that is used in its classical definition and actually establish the equivalence of the two definitions of the determinant. The determinant can be defined yet again in a third way—using the König digraph—a fact that will be useful later in the book. A special determinantal formula, derived in Section 4.3, will be used in Chapter 7. Section 4.5 describes the Laplace development of a determinant.

4.1 Definition of the Determinant

A digraph with m vertices and m edges is called *a cycle digraph*, or, more simply, a *cycle*, provided its vertices can be numbered as $1, 2, \ldots, m$ so that its set of m edges consists of edges from vertex i to vertex $i+1$, $(i = 1, 2, \ldots, m-1)$ and an edge from vertex m to vertex 1. Let D be a digraph whose set of vertices is V and whose set of edges is E. We recall from Chapter 1 that a *subdigraph* of

D is a digraph whose set of vertices is a subset U of V and whose set of edges is a subset F of the set E_U of those edges of D that join vertices in U. Thus, to form a subdigraph of D, we choose some of the vertices (possibly all of them) and some of the edges between these vertices (again, possibly all of them). If $U = V$, then we have a *spanning subdigraph* of D. If $F = E_U$, then we have an *induced subdigraph (on the set U)*.

A *linear subdigraph* of D is a spanning subdigraph of D in which each vertex has indegree 1 and outdegree 1 (i.e., exactly one edge into each vertex and exactly one (possibly the same) out of each vertex. Thus a linear subdigraph consists of a spanning collection of pairwise vertex-disjoint cycles. In Figure 4.1. a digraph D is drawn along with its three linear subdigraphs L_1, L_2, L_3.

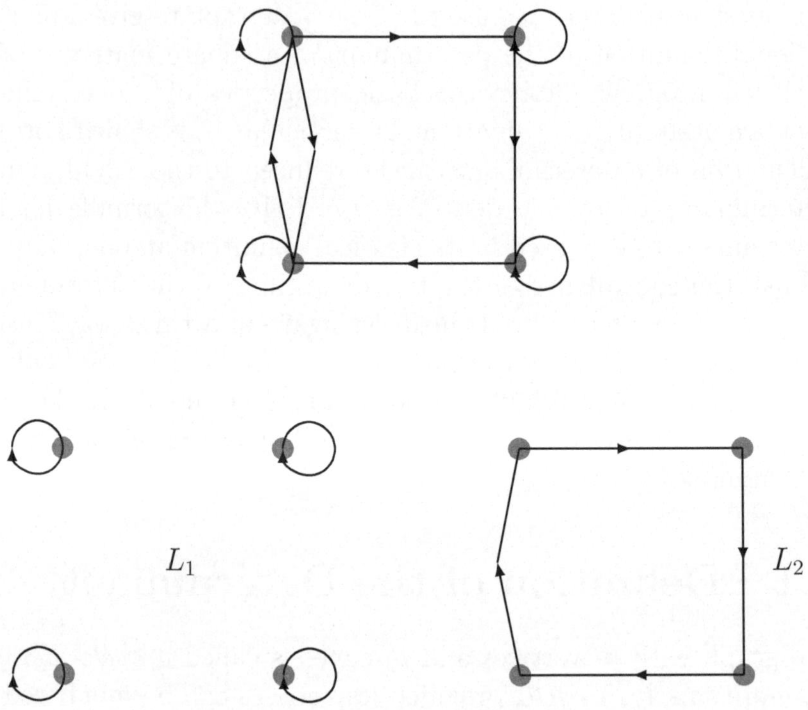

Figure 4.1

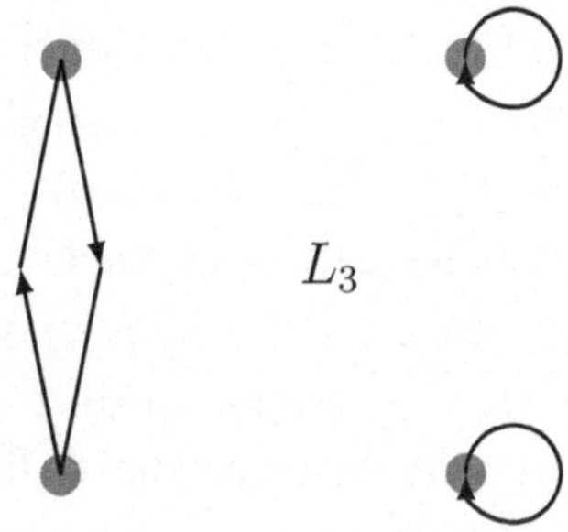

Figure 4.1 (continued)

Let $A = [a_{ij}]$ be a square matrix of order n. We have already associated two weighted digraphs with A, the König digraph $G(A)$ and the digraph $D(A)$. We now associate a third weighted digraph $D^*(A)$, which is nothing more than the digraph $D(A^T)$ associated with the transpose A^T of A. Thus $D^*(A)$ has n vertices $1, 2, \ldots, n$, and for each i, j there exists an edge from vertex j to vertex i of weight a_{ij} The elements of the main diagonal of A correspond in $D^*(A)$ to loops of $D^*(A)$ as they do in $D(A)$. The digraph $D^*(A)$ is called the *Coates digraph* of the matrix A.[1]

Let L be a linear subdigraph of the digraph $D^*(A)$. The product of the weights of the edges of L is the *weight* $w(L)$ of L. The number of cycles contained in L is denoted by $c(L)$. By $\mathcal{L}(A)$ we mean the set of all linear subdigraphs L of the Coates digraph $D^*(A)$.

Definition 4.1.1 Let $A = [a_{ij}]$ be a square matrix of order n. The *determinant* of A is the number $\det A$ defined by the sum

$$\det A = (-1)^n \sum_{L \in \mathcal{L}(A)} (-1)^{c(L)} w(L), \tag{4.1}$$

where the summation extends over all linear subdigraphs L of the digraph $D^*(A)$. Since $(-1)^{n+c(L)} = (-1)^{n-c(L)}$, another way to write (4.1) is

$$\det A = \sum_{L \in \mathcal{L}(A)} (-1)^{n-c(L)} w(L). \tag{4.2}$$

□

[1]We are using $D^*(A)$ rather than $D(A)$—that is, associating a_{ij} with the edge from j to i rather than the edge from i to j—because it aids in our later discussion.

The digraph $D^*(A)$ is named the Coates digraph and formula (4.1) is called the *Coates formula*, because Coates introduced these in [13] in developing a procedure for treating systems of linear algebraic equations (here described in Section 6.3). It is very hard to establish who first came to the idea of such a graphical interpretation of a determinant (see the Coda for some bibliographical data). Perhaps, as presented here, the definition of the determinant could be called the Harary–Coates definition since Harary, referring to Coates, has proposed in [45] that this formula, in a somewhat changed form, could be taken as the definition of the determinant. The standard definition is given in Section 4.4, and it is equivalent to the Harary–Coates definition.

Example 4.1.2 Let

$$A = \begin{bmatrix} a_{11} & a_{12} \\ a_{21} & a_{22} \end{bmatrix}.$$

Then $D^*(A)$ has two linear subdigraphs. One consists of the loops at the two vertices (so two cycles of length 1) and has weight $a_{11}a_{22}$; the other is a cycle with two vertices and has weight $a_{12}a_{21}$. Hence, using (4.1),

$$\det A = (-1)^2((-1)^2 a_{11}a_{22} + (-1)^1 a_{12}a_{21}) = a_{11}a_{22} - a_{12}a_{21}.$$

Now let

$$A = \begin{bmatrix} a_{11} & a_{12} & a_{13} \\ a_{21} & a_{22} & a_{23} \\ a_{31} & a_{32} & a_{33} \end{bmatrix}.$$

The digraph $D^*(A)$ is depicted in Figure 4.2 along with its six linear subdigraphs $L_1, L_2, \ldots, L_6$.

Applying formula (4.2) for the determinant, we get

$$\begin{aligned} \det A &= (-1)^{3-3} a_{11}a_{22}a_{33} + (-1)^{3-1} a_{12}a_{31}a_{23} + \\ &\quad (-1)^{3-1} a_{21}a_{32}a_{13} + (-1)^{3-2} a_{11}a_{23}a_{32} + \\ &\quad (-1)^{3-2} a_{22}a_{13}a_{31} + (-1)^{3-2} a_{33}a_{12}a_{21} \\ &= a_{11}a_{22}a_{33} + a_{12}a_{31}a_{23} + a_{21}a_{32}a_{13} - \\ &\quad a_{11}a_{23}a_{32} - a_{22}a_{13}a_{31} - a_{33}a_{12}a_{21}. \end{aligned}$$

From formula (4.3) we see that the determinant of a matrix of order 3 is an algebraic sum of six products with three factors, each taken according to the scheme in Figure 4.3. The + sign is ascribed to the product of elements lying on the diagonal a_{11}, a_{22}, a_{33} and to the products of elements from vertices of two triangles having one side parallel to this diagonal. The − sign is ascribed to the product of elements lying on the diagonal a_{31}, a_{22}, a_{13} and to the products of elements from vertices of two triangles having one side parallel to this diagonal. □

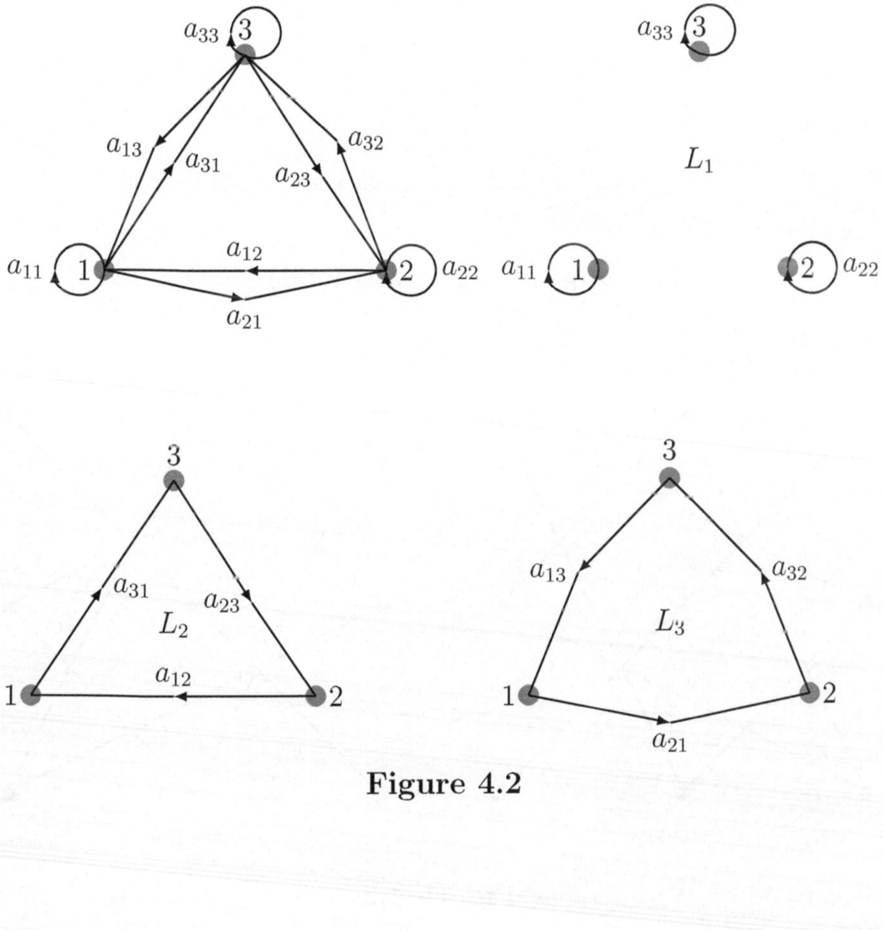

Figure 4.2

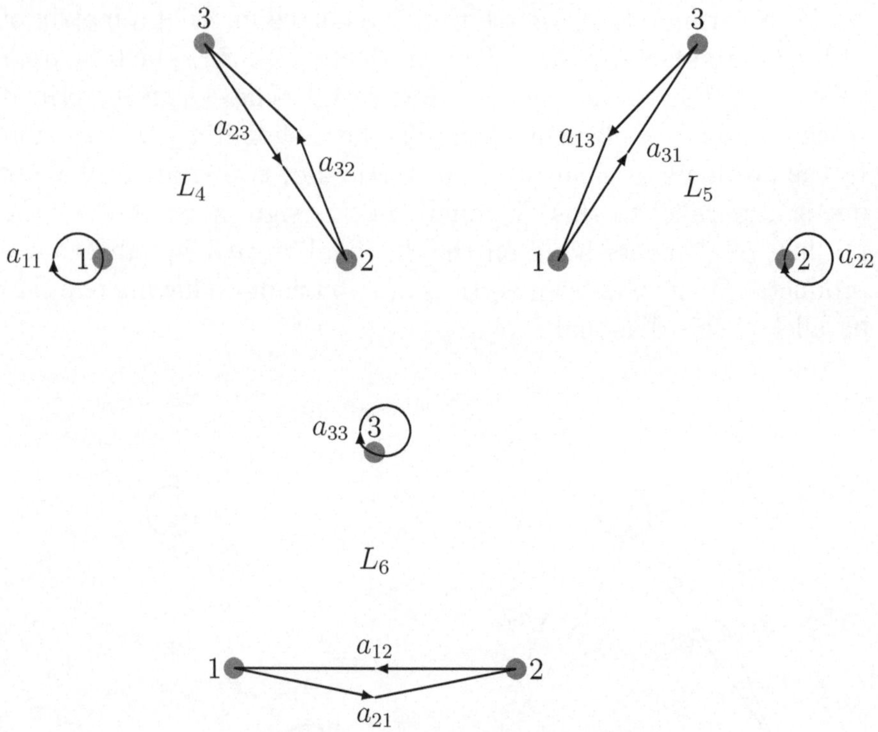

Figure 4.2 (continued)

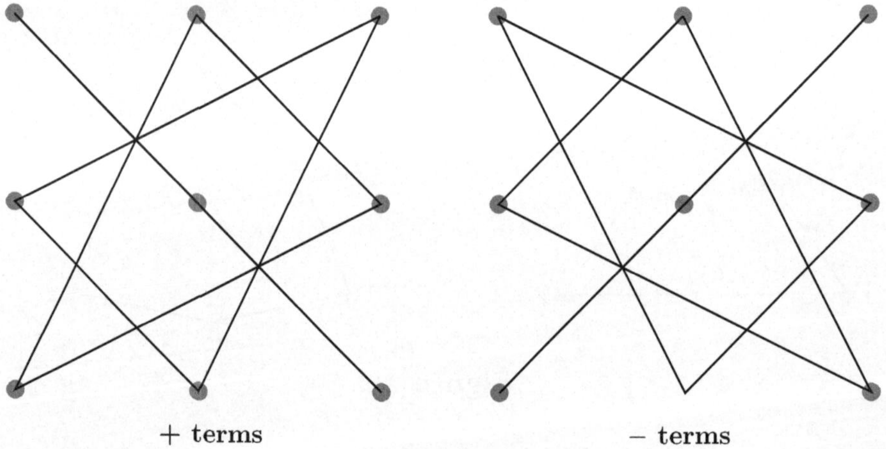

Figure 4.3

As with the digraphs $G(A)$ and $D(A)$, we adopt the convention that edges of $D^*(A)$ of weight zero are not, in general, drawn. The advantage is that, with zeros present in a matrix A, certain linear subdigraphs are removed from $\mathcal{L}(A)$, namely, those that have weight zero and thus those that do not affect the value of the determinant of A. Thus, in calculating the determinant of a matrix $A = [a_{ij}]$ of order 3, if $a_{12} = 0$, then the linear subdigraphs L_2 and L_5 of weight zero do not appear in the calculation. When a matrix A has a lot of zeros occuring in a structured way, it may be possible to easily calculate the determinant.

Example 4.1.3 In calculating the determinant of the matrix

$$A_1 = \begin{bmatrix} a_1 & 0 & \cdots & 0 \\ 0 & a_2 & \cdots & 0 \\ \vdots & \vdots & \ddots & \vdots \\ 0 & 0 & \cdots & a_n \end{bmatrix},$$

whose only nonzero elements occur on the diagonal, we see that $D^*(A_1)$ has only one linear subdigraph, namely, itself, and it is drawn in Figure 4.4.

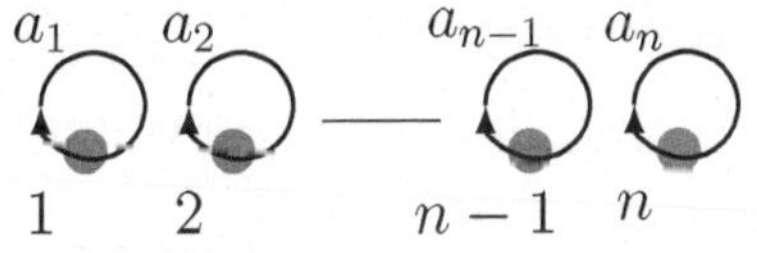

Figure 4.4

Hence we get

$$\det A_1 = (-1)^n\,(-1)^n a_{11}a_{22}\ldots a_{nn} = a_{11}a_{22}\ldots a_{nn}.$$

In calculating the determinant of the matrix

$$A_2 = \begin{bmatrix} 0 & 0 & \cdots & 0 & a_{1n} \\ 0 & 0 & \cdots & a_{2,n-1} & 0 \\ \vdots & \vdots & \vdots & \vdots & \vdots \\ a_{n1} & 0 & \cdots & 0 & 0 \end{bmatrix},$$

whose only nonzero elements are $a_{1n}, a_{2,n-2}, \ldots, a_{n1}$, we see again that $D^*(A_2)$ has only itself as a linear subdigraph and it is drawn in Figure 4.5.

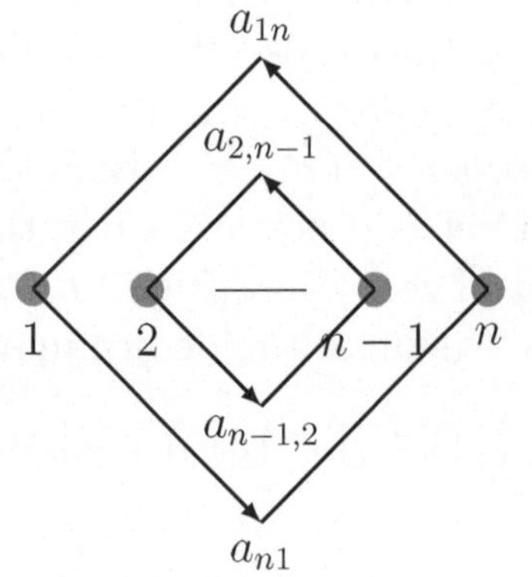

Figure 4.5

This linear subdigraph has $\lfloor (n+1)/2 \rfloor$ cycles (when n is odd, one is a loop; the others are cycles of length 2). Thus we get

$$\det A_2 = (-1)^{n-\lfloor \frac{n+1}{2} \rfloor} a_{1n} a_{2,n-1} \ldots a_{n1}.$$

□

Example 4.1.4 We calculate the determinant of the matrix

$$A = \begin{bmatrix} b & a & 0 & 0 \\ c & b & a & 0 \\ 0 & c & b & a \\ 0 & 0 & c & b \end{bmatrix}.$$

The corresponding digraph $D^*(A)$ is represented in Figure 4.6.

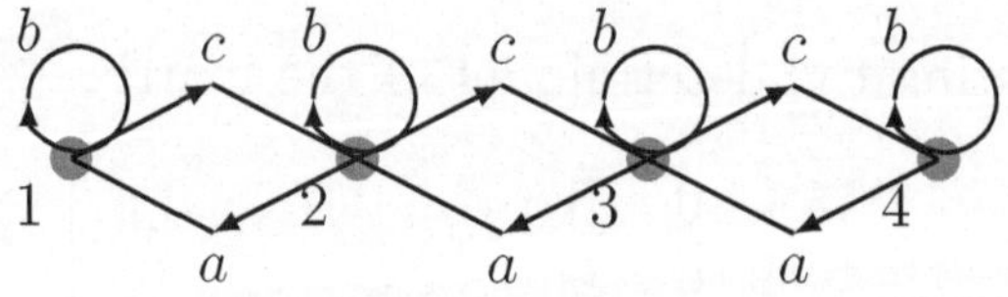

Figure 4.6

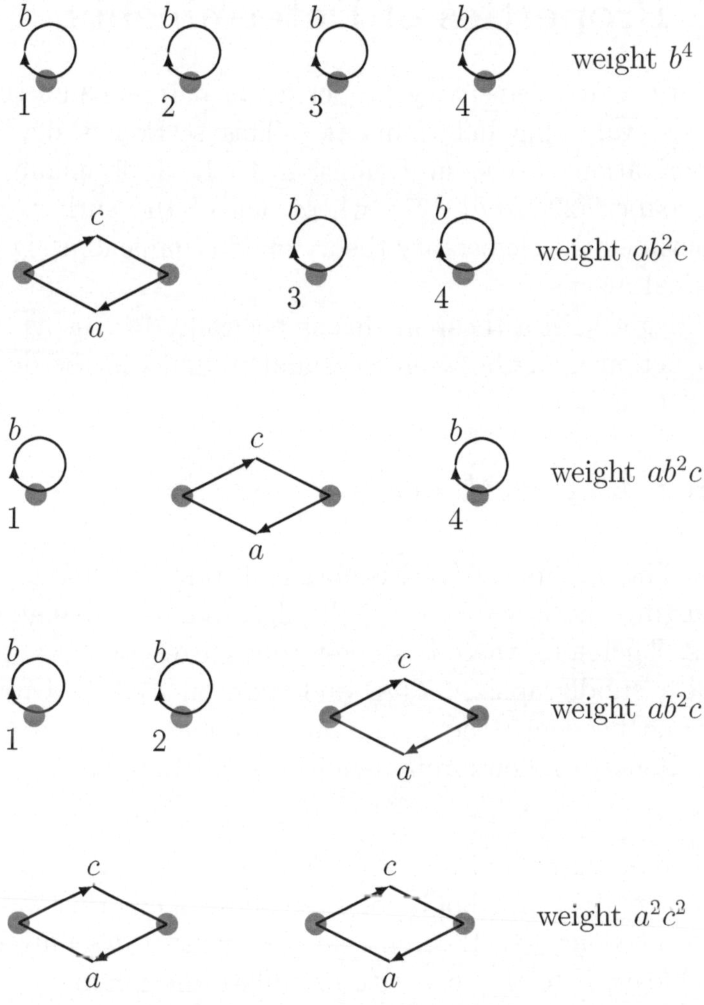

Figure 4.7

In Figure 4.7, all linear subdigraphs are given together with the corresponding weights. Therefore we have

$$\begin{aligned}\det A &= (-1)^4\left((-1)^4 b^4 + 3(-1)^3 ab^2c + (-1)^2 a^2c^2\right)\\ &= b^4 - 3ab^2c + a^2c^2.\end{aligned}$$

□

4.2 Properties of Determinants

There are many elementary properties of determinants that are useful in evaluating determinants. This section is devoted to their derivations from our Definition 4.1.1. It is mainly based on the paper [22] from 1975 where one of the authors of this book outlined the elementary theory of determinants using graph-theoretical means.

We begin with a theorem that is basically obvious as there is no distinction made between rows and columns in the definition of the determinant.

Theorem 4.2.1 $\det A^T = \det A$.

Proof. The digraph $D^*(A^T)$ is obtained from the digraph $D^*(A)$ by changing the orientation of all edges but not changing their weights. Therefore, there is a one-to-one correspondence between the linear subdigraphs in $\mathcal{L}(A)$ and those in $\mathcal{L}(A^T)$. Under this correspondence both the weight and number of cycles are preserved. Hence it follows from definition 4.1.1 that $\det A = \det A^T$. □

Theorem 4.2.1 implies that every statement that holds for the rows of a matrix also holds for the columns. In this way every theorem becomes two theorems and, in general, we only present one and leave it to the reader to formulate the other.

Theorem 4.2.2 *If each element of some row of a matrix is multiplied by c, then the determinant is also multiplied by c.*

Proof. Let each element of row i of A be multiplied by c, resulting in a matrix B. Then $D^*(A)$ and $D^*(B)$ differ only in that the weight of each edge going into vertex i is mutliplied by c in B. Each linear subdigraph contains exactly one edge going into vertex i. Hence the weight of each linear subgraph in $\mathcal{L}(B)$ is c times the weight of the corresponding linear subdigraph in $\mathcal{L}(A)$. Using the definition of the determinant, we see that $\det B = c \det A$. □

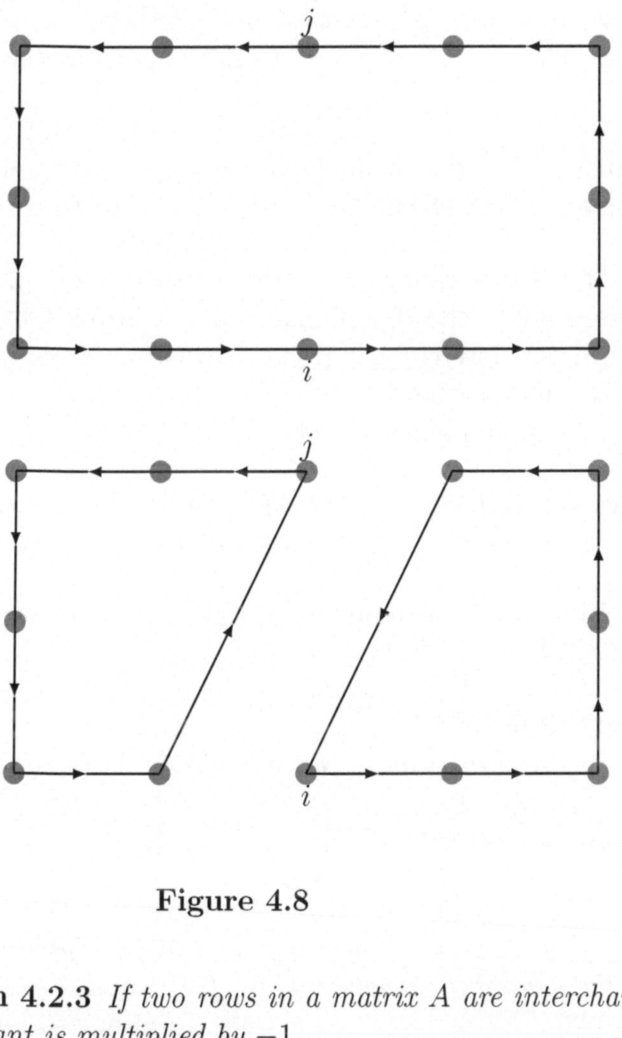

Figure 4.8

Theorem 4.2.3 *If two rows in a matrix A are interchanged, the determinant is multiplied by* -1.

Proof. Let rows i and j of the matrix A, where $i \neq j$, be interchanged, resulting in a matrix B. Then $D^*(B)$ is obtained from $D^*(A)$ by changing each edge going into vertex i into an edge going into vertex j, keeping the same weight, and viceversa. This establishes a one-to-one correspondence between the linear subdigraphs L in $\mathcal{L}(A)$ and the linear subdigraphs L' in $\mathcal{L}(B)$ that preserves the weight. However, as illustrated in Figure 4.8, the number of cycles is either increased or decreased by 1. More precisely, the number of cycles is increased by 1 if vertices i and j belong to the same cycle in L, and is decreased by 1 if they belong to different cycles. Thus $(-1)^{c(L')} = -1 \cdot (-1)^{c(L)}$, and the theorem now follows. □

Theorem 4.2.4 *If the elements of a row of a matrix A are equal to the corresponding elements of a different row, then* $\det A = 0$.

Proof. If we interchange the two identical rows of A, then. by Theorem 4.2.3, the determinant gets multiplied by -1. On the other hand, interchanging these two rows does not change the matrix. and so the determinant stays the same. Therefore, $\det A = -\det A$, implying that $\det A = 0$. □

Corollary 4.2.5 *If the elements of a row of matrix A are proportional to the elements of a different row, then* $\det A = 0$.

Proof. This corollary is an immediate consequence Theorems 4.2.2 and 4.2.4. □

Theorem 4.2.6 *Let i be a fixed integer with $1 \leq i \leq n$. Suppose that row i of A is the sum of two other rows in the sense that*

$$a_{ij} = a_{ij}^{(1)} + a_{ij}^{(2)}, \quad (1 \leq j \leq n).$$

Let $A^{(1)}$ and $A^{(2)}$ be the matrices obtained from A by replacing the element a_{ij} of row i of A with $a_{ij}^{(1)}$ and $a_{ij}^{(2)}$, respectively. Then $\det A$ *is the sum of the determinants of $A^{(1)}$ and $A^{(2)}$:*

$$\det A = \det A^{(1)} + \det A^{(2)}.$$

More generally, if row i of A is the sum of p other rows in the sense that

$$a_{ij} = a_{ij}^{(1)} + a_{ij}^{(2)} + \cdots + a_{ij}^{(p)} \quad (1 \leq j \leq n),$$

and $A^{(k)}$ is the matrix obtained fom A by replacing the elements a_{ij} in row i with $a_{ij}^{(k)}$, $(1 \leq k \leq p)$, then

$$\det A = \det A^{(1)} + \det A^{(2)} + \cdots + \det A^{(k)}.$$

Proof. As in the proof of Theorem 4.2.2, each linear subdigraph of $D^*(A)$ contains exactly one edge going into vertex i. Thus the weight of each linear subgraph L in $\mathcal{L}(A)$ contains a factor $a_{ij}^{(1)} + a_{ij}^{(2)}$ for exactly one j. This implies that there is a one-to-one correspondence between the linear subdigraphs L in $\mathcal{L}(A)$ and all pairs $L^{(1)}, L^{(2)}$ consisting of a linear subdigraph $L^{(1)}$ of $\mathcal{L}(A^{(1)})$ and a linear subdigraph $L^{(2)}$ of $\mathcal{L}(A^{(2)})$. Moreover, in this correspondence,

$$w(L) = w(L^{(1)}) + w(L^{(2)}).$$

Using Definition 4.1.1 we now compute that

$$\det A = \det A^{(1)} + \det A^{(2)}.$$

The theorem in its full generality now follows easily by induction. □

Theorem 4.2.7 *The determinant of a matrix is unchanged if the elements of some row are multiplied by a number and added to a different row.*

Proof. This theorem is an immediate upon first applying Theorem 4.2.6 and then applying Theorem 4.2.4. □

Let $v^{(1)}, v^{(2)}, \ldots, v^{(p)}$ be 1 by n row vectors (or n by 1 column vectors), and let $c_1, c_2, \ldots, c_p$ be numbers. Recall that

$$c_1 v^{(1)} + c_2 v^{(2)} + \cdots + c_p v^{(p)}$$

is a *linear combination* of $v^{(1)}, v^{(2)}, \ldots, v^{(p)}$.

Theorem 4.2.8 *Let $A = [a_{ij}]$ be a matrix of order n. Assume that some row of A is a linear combination of its other rows. Then* $\det A = 0$.

Proof. Assume, for instance, that row 1 of A is a linear combination of rows $2, 3, \ldots, n$. It follows from Theorem 4.2.6 that $\det A$ can be written as a sum of determinants of matrices whose first row is proportional to some other row of A. By Corollary 4.2.5, each of these determinants equals zero and thus $\det A = 0$. □

Example 4.2.9 Let

$$A = \begin{bmatrix} 7 & 2 & 19 \\ 1 & -2 & 3 \\ 2 & 4 & 5 \end{bmatrix}.$$

Because row 1 is three times the second row plus two times the third row, $\det A = 0$ by Theorem 4.2.8. □

Definition 4.2.10 Let $A = [a_{ij}]$ be an m by n matrix. Let $K = \{i_1, i_2, \ldots, i_k\}$ be a set of k elements with $K \subseteq \{1, 2, \ldots, m\}$, and let $L = \{j_1 . j_2, \ldots, j_l\}$ be a set of l elements with $L \subseteq \{1, 2, \ldots, n\}$. The sets K and L designate a collection of row indices and column indices, respectively, of the matrix A, and the k by l submatrix determined by them is

$$A[K, L] = \begin{bmatrix} a_{i_1 j_1} & a_{i_1 j_2} & \cdots & a_{i_1 j_l} \\ a_{i_2 j_1} & a_{i_2 j_2} & \cdots & a_{i_2 j_l} \\ \vdots & \vdots & \ddots & \vdots \\ a_{i_k j_1} & a_{i_k j_2} & \cdots & a_{i_k j_l} \end{bmatrix}.$$

If $L = K$, then $A[K, K]$ is a *principal submatrix* of A, sometimes denoted more simply as $A[K]$.

The determinant of a square submatrix of A is called a *minor* of A. Thus a minor of A equals $\det A[K, L]$, where $|K| = |L|$. If, in addition, $K = L$, then $\det A[K]$ is a *principal minor of A*.

Now assume that A is a square matrix of order n. Let i and j be integers with $1 \leq i, j \leq n$. Let A_{ij} be the submatrix of A of order $n - 1$ obtained by striking out row i and column j of A (thus, in the above notation, $A_{ij} = A[K, L]$, where $K = \{1, 2, \ldots, i - 1, i + 1, \cdots, n\}$ and $L = \{1, \cdots, j - 1, j + 1, \cdots, n\}$). The *cofactor* (or *algebraic complement*) A_{ij} of the element a_{ij} of the matrix A is given by

$$\alpha_{ij} = (-1)^{i+j} \det A_{ij}.$$

Note that the matrix A_{ij}, and hence the cofactor α_{ij}, do not depend on any of the elements in row i and column j of A. □

We now obtain a recursive formula for the determinant by showing how the determinant of a matrix of order n can be evaluated in terms of the determinants of n matrices of order $n-1$.

Theorem 4.2.11 *The determinant of a matrix $A = [a_{ij}]$ of order n can be evaluated by developing along row i as follows:*

$$\det A = \sum_{j=1}^{n} a_{ij}\alpha_{ij} = \sum_{j=1}^{n}(-1)^{i+j} a_{ij} \det A_{ij} \quad (i = 1, 2, \ldots, n).$$

It can also be evaluated by developing along column j:

$$\det A = \sum_{i=1}^{n} a_{ij}\alpha_{ij} = \sum_{i=1}^{n}(-1)^{i+j} a_{ij} \det A_{ij} \quad (j = 1, 2, \ldots, n).$$

Proof. Because of Theorem 4.2.1, it suffices to prove the formula for development along row i. In addition, by Theorem 4.2.3, it suffices to prove the theorem for $i = n$. This is because by successively interchanging row i with rows $i+1, i+2, \ldots, n$, we obtain a matrix $B = [b_{ij}]$, where in B the rows of A are in the order $1, \ldots, i-1, i+1, \ldots, n, i$. With these interchanges we have $\det B = (-1)^{n-i} \det A$. Moreover, using the notation in Definition 4.2.10, we have $b_{nj} = a_{ij}$ and $B_{nj} = A_{ij}$. Hence, developing the determinant of B along its row n, we get

$$\begin{aligned}
\det A &= (-1)^{n-i} \det B \\
&= (-1)^{n-i} \sum_{j=1}^{n}(-1)^{n+j} b_{nj} \det B_{nj} \\
&= (-1)^{n-i} \sum_{j=1}^{n}(-1)^{n+j} a_{ij} \det A_{ij} \\
&= \sum_{j=1}^{n}(-1)^{-i+j} a_{ij} \det A_{ij} \\
&= \sum_{j=1}^{n}(-1)^{i+j} a_{ij} \det A_{ij}.
\end{aligned}$$

So we need only establish the case $i = n$, and we proceed to do so.

Each term in the sum in Definition 4.1.1 for the determinant contains one element from each row of A, since in a linear subdigraph there is exactly one edge coming into each vertex. Similarly, each term contains one element from each column. Hence we may write

$$\det A = \sum_{j=1}^{n} a_{nj}\beta_j, \tag{4.3}$$

where β_j does not depend on any of the elements in row n and column j of A, $(j = 1, 2, \ldots, n)$. The terms in this summation correspond to a partition of the linear subdigraphs in $\mathcal{L}(A)$ into $\mathcal{L}_1(A), \mathcal{L}_2(A) \ldots, \mathcal{L}_n(A)$, where $\mathcal{L}_j(A)$ consists of those linear subdigraphs where the edge from vertex j goes to vertex n $(j = 1, 2, \ldots, n)$. The linear subdigraphs L in $\mathcal{L}_n(A)$ contain a loop at vertex n. Deleting that loop (so a cycle) from L, we get a linear subdigraph L' in $\mathcal{L}(A_{nn})$, where $w(L) = a_{nn}w(L')$ and $c(L) = c(L') + 1$, and so $(-1)^{c(L)} = -(-1)^{c(L')}$. Therefore

$$\begin{aligned}
\beta_n &= (-1)^n(-1) \sum_{L' \in \mathcal{L}(A_{nn})} (-1)^{c(L')} w(L') \\
&= (-1)^{n-1} \sum_{L' \in \mathcal{L}(A_{nn})} (-1)^{c(L')} w(L') \\
&= \det A_{nn} \\
&= (-1)^{n+n} \det A_{nn} = \alpha_{nn}.
\end{aligned}$$

We now consider the coefficient β_j of a_{nj} in (4.3) where $1 \le j < n$. By successively interchanging column j with columns $j+1, j+2, \ldots, n$ we obtain a matrix $C = [c_{ij}]$, where in C the columns of A are in the order $1, \ldots, j-1, j+1, \ldots, n, j$. We have $\det C = (-1)^{n-j} \det A$. Using the notation in Definition 4.2.10, we have $c_{nn} = a_{nj}$ and $C_{nn} = A_{nj}$. It follows from what we have proved in the preceding paragraph that

$$\beta_j = (-1)^{n-j} \det C_{nn} = (-1)^{n+j} \det C_{nj} = (-1)^{n+j} \det A_{nj} = \alpha_{nj}.$$

Therefore we have

$$\det A = \sum_{j=1}^{n} a_{nj}\alpha_{nj},$$

as desired. □

Example 4.2.12 Let

$$A = \begin{bmatrix} 1 & 0 & 2 & 1 \\ 0 & 3 & 2 & 0 \\ 2 & 1 & 3 & 4 \\ 1 & 3 & 2 & 0 \end{bmatrix}.$$

Developing the determinant along row 2 and taking into account the two 0's in row 2, we get that

$$\begin{aligned} \det A &= (-1)^4 \cdot 3 \cdot \det \begin{bmatrix} 1 & 2 & 1 \\ 2 & 3 & 4 \\ 1 & 2 & 0 \end{bmatrix} + (-1)^5 \cdot 2 \cdot \det \begin{bmatrix} 1 & 0 & 1 \\ 2 & 1 & 4 \\ 1 & 3 & 0 \end{bmatrix} \\ &= (-1)^4 \cdot 3 \cdot (1) + (-1)^5 \cdot 2 \cdot (-7) \\ &= 3 + 14 = 17. \end{aligned}$$

The two determinants of order 3 can be computed either by the formula given in Example 4.1.2 or by further determinant development. □

We conclude this section by deriving two more important properties of the determinant.

Theorem 4.2.13 *Let*

$$A = \begin{bmatrix} A_1 & 0 \\ B & A_2 \end{bmatrix},$$

where A_1 and A_2 are square submatrices of A. Then

$$\det A = \det A_1 \det A_2.$$

In particular, the determinant of A does not depend on B.

Proof. Let A, A_1, A_2 be matrices of orders n, n_1, n_2 respectively, where $n = n_1 + n_2$. The digraph $D = D^*(A)$ is formed from the digraphs $D_1 = D^*(A_1)$ and $D_2 = D^*(A_2)$ by including some edges that go from the vertices of D_1 to the vertices of D_2. These edges correspond to the nonzero entries of B. Because no edges go from D_2 to D_1, no cycle of D contains edges corresponding to the nonzero entries of B. Therefore, each linear subdigraph L of D consists of the union of a linear subdigraph L_1 of D_1 and a linear subdigraph L_2 of D_2. Moreover, every such union gives a linear subdigraph of D. We thus have

$$c(L) = c(L_1) + c(L_2) \text{ and } w(L) = w(L_1)w(L_2),$$

and

$$\begin{aligned}
\det A &= (-1)^n \sum_{L\in\mathcal{L}(A)} (-1)^{c(L)} w(L) \\
&= (-1)^{n_1+n_2} \sum_{L_1\in\mathcal{L}(A_1)} \sum_{L_2\in\mathcal{L}(A_2)} (-1)^{c(L_1)+c(L_2)} w(L_1)w(L_2) \\
&= (-1)^{n_1} \sum_{L_1\in\mathcal{L}(A_1)} (-1)^{c(L_1)} w(L_1) \cdot (-1)^{n_2} \sum_{L_2\in\mathcal{L}(A_2)} (-1)^{c(L_2)} w(L_2) \\
&= \det A_1 \det A_2.
\end{aligned}$$

□

Theorem 4.2.13 can be used to show that the determinant is a multiplicative function.

Theorem 4.2.14 *If A and B are square matrices of the same order n, then*

$$\det AB = \det A \det B. \tag{4.4}$$

Proof. From Theorem 4.2.13 we get

$$\det \begin{bmatrix} A & 0 \\ -I_n & B \end{bmatrix} = \det A \det B. \tag{4.5}$$

We multiply column 1 of the matrix of order $2n$ in (4.5) by b_{11}, column 2 by b_{21}, …, and column n by b_{n1}, and add each of them to the column $n+1$. Furthermore, we multiply column 1 by b_{12}, column 2 by b_{22}, …, and column n by b_{n2}, and add each of them to the column $n+2$. Continuing like and using the fact that by Theorem 4.2.7 the determinant is unchanged, we obtain

$$\det \begin{bmatrix} A & 0 \\ -I_n & B \end{bmatrix} = \det \begin{bmatrix} A & AB \\ -I_n & 0 \end{bmatrix}. \tag{4.6}$$

Now by n interchanges of pairs of columns (1 and $n+1$, 2 and $n+2$, $\ldots$, n and $2n$) we get by Theorems 4.2.3 and 4.2.13 that

$$\begin{aligned} \det \begin{bmatrix} A & AB \\ -I_n & 0 \end{bmatrix} &= (-1)^n \det \begin{bmatrix} AB & A \\ 0 & -I_n \end{bmatrix} \\ &= (-1)^n \det(AB) \det(-I_n) \\ &= (-1)^n \det AB\ (-1)^n = \det AB. \end{aligned} \tag{4.7}$$

Combining (4.5), (4.6), and (4.7), we obtain (4.4). □

Theorem 4.2.14 asserts that the determinant of a product of square matrices of the same order equals the product of the determinants of each of the two matrices. The product AB of two nonsquare matrices A and B may be a square matrix and then will have a determinant, even though neither factor does. In fact, this happens exactly when A is an m by n matrix and B is an n by m matrix for some integers m and n. In this situation, AB is a square matrix of order m, and it is natural to ask whether or not there is a formula for $\det(AB)$ that generalizes the product rule of Theorem 4.2.14. Such a formula exists, and it is called the Binet–Cauchy formula.

Example 4.2.15 Let

$$A = \begin{bmatrix} 1 \\ b \\ c \end{bmatrix} \text{ and } B = \begin{bmatrix} x & y & z \end{bmatrix}.$$

Then

$$AB = \begin{bmatrix} x & y & z \\ bx & by & bz \\ cx & cy & cz \end{bmatrix}.$$

Applying Corollary 4.2.5, we see that since rows 1 and 2 are proportional (as are rows 1 and 3), $\det(AB) = 0$.

This example is a special case of a more general situation. Let

$$A = \begin{bmatrix} a_1 \\ a_2 \\ \vdots \\ a_m \end{bmatrix} \text{ and } B = \begin{bmatrix} b_1 & b_2 & \cdots & b_m \end{bmatrix}$$

be m by 1 and 1 by m matrices, respectively. Then

$$AB = \begin{bmatrix} a_1b_1 & a_1b_2 & \cdots & a_1b_m \\ a_2b_1 & a_2b_2 & \cdots & a_2b_m \\ \vdots & \vdots & \ddots & \vdots \\ a_mb_1 & a_mb_2 & \cdots & a_mb_m \end{bmatrix}$$

is a square matrix of order m. Each pair of rows and each pair of columns is proportional. Thus, if $m > 1$, $\det(AB) = 0$. □

Let $A = [a_{ij}]$ and $B = [b_{ij}]$. Let the columns of A be the m by 1 matrices $C_1, C_2, \ldots, C_n$, and let the rows of B be the 1 by m matrices $R_1, R_2, \ldots, R_n$. Then it follows from the definition of matrix multiplication that

$$AB = C_1R_1 + C_2R_2 + \cdots + C_nR_n. \tag{4.8}$$

This is because the element in position (i, j) of AB is $a_{i1}b_{1j} + a_{i2}b_{2j} + \cdots + a_{in}b_{nj}$, and this is the sum of the elements in position (i, j) of the matrices $C_1R_1, C_2R_2, \ldots, C_nR_n$.

The next theorem contains the *Binet–Cauchy formula.*

Theorem 4.2.16 *Let A and B be m by n and n by m matrices, respectively. If $m > n$, then $\det(AB) = 0$. If $m \leq n$, then*

$$\det(AB) = \sum_K \det A[\{1, 2, \ldots, m\}, K] \det B[K, \{1, 2, \ldots, m\}], \tag{4.9}$$

where the summation extends over all subsets K of $\{1, 2, \ldots, n\}$ of cardinality m.

Proof. First assume that $m > n$. Then, from (4.8), we conclude that the columns of AB are linear combinations of the n columns of B. Applying Theorem 4.2.8 and 4.2.2, we see that $\det A$ is a sum of multiples of determinants of matrices, each of whose m rows is one of the n rows of B. Because $m > n$, each of these matrices has two equal rows and so by Theorem 4.2.4, $\det A = 0$.

Now let $m \leq n$. Let C be the matrix of order $m + n$ defined by

$$C = \begin{bmatrix} A & O \\ -I_n & B \end{bmatrix}.$$

Similar to the proof of Theorem 4.2.14, we obtain by interchanging each of the last m columns with each of the first n columns in turn[2], that

$$\det C = \det \begin{bmatrix} A & AB \\ -I_n & B \end{bmatrix}. \tag{4.10}$$

Similar to the proof of Theorem 4.2.14, we also obtain by interchanging each of the last m columns with each of the first n columns in turn[3] that

$$\det C = \det \begin{bmatrix} A & AB \\ -I_n & O \end{bmatrix} = (-1)^{mn} \det \begin{bmatrix} AB & A \\ O & -I_n \end{bmatrix}$$
$$= (-1)^{mn+n} \det(AB). \tag{4.11}$$

Since A and B are, in general, nonsquare matrices, we cannot invoke Theorem 4.2.13 to conclude that the first determinant in (4.11) equals $\det(A)\det(B)$.

Consider the Coates digraph $D^*(C)$ with vertices $\{1, 2, \ldots, m+n\}$, and a linear subdigraph L in $\mathcal{L}(C)$ with nonzero weight. In order that L have nonzero weight, exactly $n-m$ edges must correspond to -1's on the diagonal of $-I_n$. This implies that there is a subset K of $\{1, 2, \ldots, n\}$ of cardinality m such that edges in L from these vertices go to vertices $1, 2, \ldots, m$ (and thus their weights comes from elements of A). It then follows that the edges in L from the last m vertices (the vertices $n+1, n+2, \ldots, m+n$) go to the vertices corresponding to the rows of B whose ordinal numbers are also in K.

Let $\mathcal{L}_K(C)$ be the subset of $\mathcal{L}(C)$ consisting of all those linear subdigraphs for which the edges from the vertices in K go to the vertices $\{1, 2, \ldots, m\}$. We then have

$$\det(C) = (-1)^{m+n} \sum_{K \subseteq \{1,2,\ldots,n\}, |K|=m} w(\mathcal{L}_K(C)),$$

where

$$w(\mathcal{L}_K(C)) = \sum_{L \in \mathcal{L}_K(C)} (-1)^{c(L)} w(L).$$

[2]Note that if we had interchanged like this in the proof of Theorem 4.2.14 we would have the sign $(-1)^{n^2}$ instead of $(-1)^n$. Since n^2 is even if and only if n is even, we have $(-1)^{n^2} = (-1)^n$.

[3]See the preceding footnote.

We may perform column interchanges so that the columns in K come first (followed by the remaining columns in A in the same relative order they appear in A) and do similar row interchanges so that the rows in K of B come first in C. This implies that $w(\mathcal{L}_K(C))$ equals

$$\det \begin{bmatrix} A[\{1,2,\ldots,m\},K] & O & O \\ O & O & B[K,\{1,2,\ldots,m\}] \\ O & -I_{n-m} & O \end{bmatrix}.$$

By interchanging $(n-m)m$ pairs of columns we see that $w(\mathcal{L}_K(C))$ equals

$$= (-1)^{m(n-m)} \det \begin{bmatrix} A[\{1,2,\ldots,m\},K] & O & O \\ O & B[K,\{1,2,\ldots,m\}] & O \\ O & O & -I_{n-m} \end{bmatrix}.$$

By two applications of Theorem 4.2.14 applied to this last determinant, we see that $w(\mathcal{L}_K(C))$ equals

$$(-1)^{m(n-m)} \det A[\{1,2,\ldots,m\},K] \det B[K,\{1,2,\ldots,m\}](-1)^{n-m}.$$

Now, using (4.10), we see that the contribution of $\mathcal{L}_K(C)$ to the determinant of C equals

$$(-1)^{mn+n+m(n-m)+(n-m)} \det A[\{1,2,\ldots,m\},K]$$
$$\det B[K,\{1,2,\ldots,m\}].$$

Because $mn+n+m(n-m)+(n-m) = 2mn+2n-m(m+1)$ is always an even number, we get (4.9). □

Example 4.2.17 Let

$$A = \begin{bmatrix} 1 & 2 & 3 \\ 4 & 5 & 6 \end{bmatrix} \text{ and } B = \begin{bmatrix} 1 & 2 \\ 3 & 4 \\ 5 & 6 \end{bmatrix}.$$

Then

$$AB = \begin{bmatrix} 22 & 28 \\ 49 & 64 \end{bmatrix}, \text{ and}$$

$$\det(AB) = (22)(64) - (28)(49) = 1408 - 1372 = 36.$$

Using the Binet–Cauchy theorem we get

$$\det(AB) = \det\begin{bmatrix} 1 & 2 \\ 4 & 5 \end{bmatrix} \det\begin{bmatrix} 1 & 2 \\ 3 & 4 \end{bmatrix} + \det\begin{bmatrix} 1 & 3 \\ 4 & 6 \end{bmatrix} \det\begin{bmatrix} 1 & 2 \\ 5 & 6 \end{bmatrix} +$$
$$\det\begin{bmatrix} 2 & 3 \\ 5 & 6 \end{bmatrix} \det\begin{bmatrix} 3 & 4 \\ 5 & 6 \end{bmatrix},$$

and so

$$\det(AB) = (-3)(-2) + (-6)(-4) + (-3)(-2) = 6 + 24 + 6 = 36.$$

□

4.3 A Special Determinant Formula

Let $A = [a_{ij}]$ be a matrix of order n, and let λ be a variable. The matrix $A + \lambda I$ is obtained from A by adding λ to each diagonal entry. Thus the weight a_{ii} of the loop at vertex i in $D^*(A)$ is replaced by $a_{ii} + \lambda$ in $D^*(A + \lambda I)$; there are no other changes in the weights. It follows from the definition of the determinant that $\det(A + \lambda I)$ is a polynomial in λ of degree n. In this section, we identify that polynomial.

Let $\mathcal{L}$ be the set of linear subdigraphs of $D^*(A + \lambda I)$. By definition,

$$\det(A + \lambda I) = (-1)^n \sum_{L \in \mathcal{L}} (-1)^{c(L)} w(L). \tag{4.12}$$

Consider a linear subdigraph L and suppose that L contains exactly k loops $(0 \le k \le n)$. Let the vertices with these loops be the vertices $i_1, i_2, \ldots, i_k$ with $1 \le i_1 < i_2 < \cdots < i_k \le n$ (if $k = 0$, there are no loops). Then the weight of L is $(a_{i_1 i_1} + \lambda)(a_{i_2 i_2} + \lambda) \cdots (a_{i_k i_k} + \lambda)\beta(L)$, where $\beta(L)$ is the product of the weights of the nonloop edges in L and thus does not depend on λ. There are 2^k terms when the product $(a_{i_1 i_1} + \lambda)(a_{i_2 i_2} + \lambda) \cdots (a_{i_k i_k} + \lambda)$ is multiplied out, and these terms are of the form

$$\lambda^p a_{j_1 j_1} a_{j_2 j_2} \cdots a_{j_{k-p} j_{k-p}},$$

where $0 \le p \le k$ and $\{j_1, j_2, \ldots, j_{k-p}\}$ is a subset of $\{i_1, i_2, \ldots, i_k\}$. Thus the weight of L satisfies

$$w(L) = \lambda^p w(L'), \tag{4.13}$$

where L' is a linear subdigraph of the digraph $D^*(A')$ (an induced subdigraph of $D^*(A)$) and where A' is a principal submatrix[4] of A of order $n - p$ obtained by striking out p rows and p columns, namely, those rows and columns whose indices belong to $\{i_1, i_2, \ldots, i_k\} \setminus \{j_1, j_2, \ldots, j_{k-p}\}$, the complement of $\{j_1, j_2, \ldots, j_{k-p}\}$ in $\{i_1, i_2, \ldots, i_k\}$. Conversely, every such linear subdigraph L' contributes a term to (4.13).

Putting this all together we obtain the determinant formula in the following theorem:

Theorem 4.3.1 *Let A be a matrix of order n. Then*

$$\det(A + \lambda I) = \sum_{p=0}^{n} \lambda^p c_{n-p},$$

where c_{n-p} equals the sum of the principal minors of order $n - p$ of A.

□

By replacing λ with $-\lambda$ in Theorem 4.3.1, and then multiplying $A - \lambda I$ by -1 to produce $\lambda I - A$, and by using the fact that $(-1)^{n+p} = (-1)^{n-p}$, we obtain the following corollary.

Corollary 4.3.2 *Let A be a matrix of order n. Then*

$$\det(A - \lambda I) = \sum_{p=0}^{n} (-1)^p \lambda^p c_{n-p},$$

where c_{n-p} equals the sum of the principal minors of order $n - p$ of A. Equivalently,

$$\det(\lambda I - A) = \sum_{p=0}^{n} (-1)^{n-p} \lambda^p c_{n-p}.$$

□

[4]Note: Principal submatrix of A, not principal submatrix of $A + \lambda I$.

In Theorem 4.3.1 and Corollary 4.3.2, because $\det A$ is the only principal minor of order n of A, the constant term c_n equals the determinant of A. The coefficient c_1 of λ^{n-1} equals the sum of the principal minors of order 1, and this is the *trace* of A. Thus the trace of A is given by

$$\operatorname{tr}(A) = a_{11} + a_{22} + \cdots + a_{nn}.$$

Example 4.3.3 Let

$$A = \begin{bmatrix} 1 & 0 & 1 \\ 1 & 1 & 1 \\ 1 & 1 & 0 \end{bmatrix}.$$

Then

$$\det(A + \lambda I) = \lambda^3 + 2\lambda^2 + (-1)\lambda + (-1) = \lambda^3 + 2\lambda^2 - \lambda - 1.$$

□

4.4 Classical Definition of the Determinant

The determinant formula in terms of linear subdigraphs of the Coates digraph, as given in Definition 4.1.1, is not the formula that is usually given initially for the determinant. The classical formula involves permutations and their signs ($\pm$). In this section we show that this classical formula is equivalent to our formula.

Let $A = [a_{ij}]$ be a square matrix of order n, and let L be a linear subgraph of $D^*(A)$. Because L contains one edge into each vertex and one edge out of each vertex, the weight of L is the product of n entries of A consisting simultaneously of one element from each row of A (of the n edges in L, exactly one of them comes into each vertex) and one entry from each column (of the n edges in L, exactly one of them comes out of each vertex). If we arrange these n entries according to increasing row indices, then we see that

$$w(L) = a_{1j_1} a_{2j_2} \cdots a_{nj_n}, \tag{4.14}$$

where $(j_1, j_2, \ldots, j_n)$ is a permutation[5] of $\{1, 2, \ldots, n\}$. Thus the edges of L are the n edges from vertex j_1 to vertex 1, from vertex j_2 to vertex 2, ..., and from vertex j_n to vertex n. Conversely, each product of n entries of A, one from each row and simultaneously one from each column (so a product as given in (4.14)), is the weight of some linear subdigraph of $D^*(A)$. In formula (4.2) for the determinant, $w(L)$ has a sign affixed in front of it, namely, $(-1)^{n-c(L)}$. Let S_n denote the set of all $n!$ permutations of $\{1, 2, \ldots, n\}$. Using our notation, we can write

$$\det A = \sum_{(j_1, j_2, \ldots, j_n) \in S_n} (-1)^{n-c(L)} a_{1j_1} a_{2j_2} \cdots a_{nj_n}.$$

What we would like to do is determine how to write the sign $(-1)^{n-c(L)}$ in terms of the corresponding permutation $(j_1, j_2, \ldots, j_n)$.

Let $\sigma = (j_1, j_2, \ldots, j_n)$ be in S_n. An *inversion* of σ is a pair k, l of integers with $1 \leq k < l \leq n$ such that $j_k > j_l$. Thus an inversion represents a pair of integers out of their natural order in σ. Let $\#(\sigma)$ equal the *number of inversions of* σ. The *sign of the permutation* σ is defined to be $(-1)^{\#(\sigma)}$. The permutation σ is an *even permutation* if it has an even number of inversions (i.e., its sign is $+1$) and is an *odd permutation* if it has an odd number of inversions (i.e., its sign is -1).

Example 4.4.1 Let $n = 6$ and let $\sigma = (5, 4, 1, 3, 6, 2)$. Then σ has inversions corresponding to the following pairs of integers out of their natural order in σ:

$$5,4;\ 5,1;\ 5,3;\ 5,2;\ 4,1;\ 4,3;\ 4,2;\ 3,2;\ 6,2.$$

Since $\#(\sigma) = 9$, σ is an odd permutation (its sign is -1). The permutation σ corresponds to a linear subdigraph L of the Coates digraph of a matrix of order 6, where L is a cycle consisting of edges from vertices 5 to 1, 1 to 3, 3 to 4, 4 to 2, 2 to 6, and 6 to 1,

Now consider the identity permutation $\iota = (1, 2, 3, 4, 5, 6)$. Then ι has no inversions and so is an even permutation (its sign is $+1$). The permutation ι corresponds to a linear subdigraph of a Coates digraph consisting of a loop at each of the six vertices.□

[5]We now write a permutation of $\{1, 2, \ldots, n\}$ as an n-tuple.

In what is to follow we refer to the decomposition into cycles of the linear subdigraph L corresponding to a permutation σ of $\{1, 2, \ldots, n\}$ as the *cycle decomposition of the permutation* σ, and we denote the number of cycles by $c(\sigma)$. Thus $c(\sigma) = c(L)$.

Lemma 4.4.2 *Let $(j_1, j_2, \ldots, j_n)$ be a permutation of $\{1, 2, \ldots, n\}$. Then $\#(\sigma)$ and $n - c(\sigma)$ have the same parity. Therefore,*

$$(-1)^{n-c(\sigma)} = (-1)^{\#(\sigma)}.$$

Proof. We prove the lemma by backwards induction on the number of cycles of σ. To get the induction started, assume that $c(\sigma) = n$, the largest possible number. Then σ is the identity permutation, $\#(\sigma) = 0$, and $c(\sigma) = n$. Hence, in this case, $n - c(\sigma) = \#(\sigma) = 0$, in particular, $n - c(\sigma)$ and $\#(\sigma)$ have the same parity.

We now assume that $c(\sigma) < n$. Then $\sigma \neq \iota$ and σ contains a cycle j_{k_1} to j_{k_2}, j_{k_2} to j_{k_3}, j_{k_3} to ... to $j_{k_{t-1}}$, $j_{k_{t-1}}$ to j_{k_t}, j_{k_t} to j_{k_1} of $t \geq 2$ elements where $k_1 < k_2 < \cdots < k_t$. There must be an inversion pair j_{k_r} and j_{k_s} with $r < s$ and $j_{k_r} > j_{k_s}$. If we interchange j_{k_r} and j_{k_s} in σ, we obtain a new permutation τ such that $\#(\tau)$ and $\#(\sigma)$ differ by an odd number and, in addition, $c(\tau) = c(\sigma) + 1$ (and so differ by an odd number). By induction, $\#(\tau)$ and $n - c(\tau)$ have the same parity, and hence so do $\#(\sigma)$ and $c(\sigma)$. □

Summarizing, we now arrive at a theorem containing the classical definition of the determinant.

Theorem 4.4.3 *Let $A = [a_{ij}]$ be a square matrix of order n. Then*

$$\det A = \sum_{(j_1,j_2,\ldots,j_n)\in S_n} (-1)^{\#(j_1,j_2,\ldots,j_n)} a_{1j_1} a_{2j_2} \cdots a_{nj_n}, \tag{4.15}$$

where the summation extends over all permutations $(j_1, j_2, \ldots, j_n)$ of te integers $1, 2, \ldots, n$.

□

Example 4.4.4 Let

$$A = \begin{bmatrix} a & b & 0 & c \\ 0 & d & e & 0 \\ 0 & 0 & f & g \\ p & 0 & 0 & q \end{bmatrix}.$$

It is straighforward to check that there are only three permutations that give nonzero terms in formula (4.15), namely, $(1,2,3,4)$ (no inversions), $(2,3,4,1)$ (3 inversions), and $(4,2,3,1)$ (5 inversions). These permutations are even, odd, and odd, respectively. Hence

$$\det A = adeg - begp - cdfp.$$

□

We conclude this section by reformulating the classical definition of the determinant in terms of the König digraph of a matrix.

Let $A = [a_{ij}]$ be a square matrix of order n. The König digraph $G(A)$ has n black vertices and n white vertices. A collection F of n edges of $G(A)$, one leaving each black vertex and one terminating at each white vertex, is a *1-factor* of $G(A)$ (see Section 1.1). The *weight* $w(F)$ *of the 1-factor* F is the product of the weights of its edges. The 1-factors of $G(A)$ are in one-to-one correspondence with the terms in the classical determinant formula (4.15). Let $(j_1, j_2, \ldots, j_n)$ be a permutation of $\{1, 2, \ldots, n\}$. To the term

$$(-1)^{\#(j_1,j_2,\ldots,j_n)} a_{1j_1} a_{2j_2} \cdots a_{nj_n}$$

in formula (4.15) we let correspond the n edges $e_1, e_2, \ldots, e_n$ of $G(A)$, where e_i is the edge from black vertex i to white vertex $j_i, (i = 1, 2, \ldots, n)$. Because $(j_1, j_2, \ldots, j_n)$ is a permutation of $\{1, 2, \ldots, n\}$, the resulting set of edges $\{e_1, e_2, \ldots, e_n\}$ is a 1-factor F of $G(A)$ and its weight is $w(F) = a_{1j_1} a_{2j_2} \cdots a_{nj_n}$. Each 1-factor of $G(A)$ arises from a permutation of $\{1, 2, \ldots, n\}$ in this way.

Let us draw the digraph $G(A)$ so that white vertex i is placed directly above black vertex i, as in Figure 4.9.

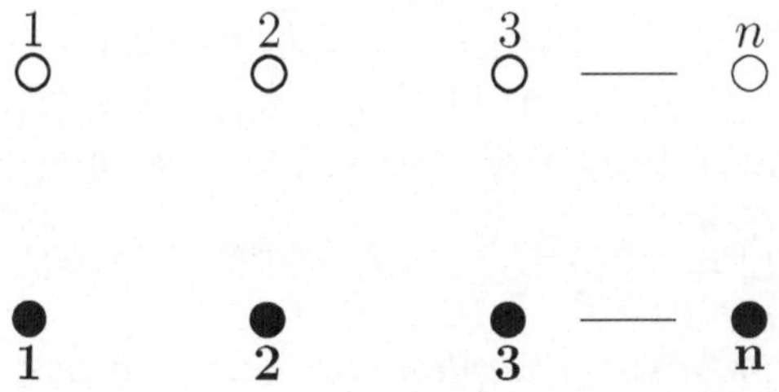

Figure 4.9

Let $q(F)$ equal the number of pairs of edges in F that intersect each other in drawing $G(A)$ in this way. Let e_k, joining black vertex k to white vertex j_k, and e_l, joining black vertex l to white vertex j_l, be two edges in F with $k < l$. Then e_k and e_l intersect exactly when $j_l > j_k$. Thus the intersections of edges of F are in one-to-one correspondence with the inversions of the permutation $(j_1, j_2, \ldots, j_n)$, and hence $q(F) = \#(j_1, j_2, \ldots, j_n)$. Let $\mathcal{F}(A)$ denote the collection of 1-factors of $G(A)$. We then have the following reformulation for the determinant of A:

$$\det A = \sum_{F \in \mathcal{F}(A)} (-1)^{q(F)} w(F). \tag{4.16}$$

4.5 Laplace Development of the Determinant

In this section we generalize the recursive formula

$$\det A = \sum_{j=1}^{n} (-1)^{i+j} a_{ij} \det A_{ij}, \quad (i = 1, 2, \ldots, n)$$

for the determinant given in Theorem 4.2.11.

Let $A = [a_{ij}]$ be a square matrix of order n. Let

$$K = \{k_1, k_2, \ldots, k_\nu\} \text{ and } L = \{l_1, l_2, \ldots, l_\nu\}$$

be subsets of $\{1, 2, \ldots, n\}$ of the same cardinality ν. Recall that $\det A[K, L]$ is a minor of A of order ν.

Let $\overline{K} = \{1, 2, \ldots, n\} \setminus K$ and $\overline{L} = \{1, 2, \ldots, n\} \setminus L$ be the complements of K and L in $\{1, 2, \ldots, n\}$, respectively. (We always assume that the indices are written in increasing order.) Then

$$\Delta[K, L] = (-1)^{k_1+k_2+\cdots+k_\nu+l_1+l_2+\cdots+l_\nu} \det A[\overline{K}, \overline{L}]$$

is the the *algebraic complement* or *cofactor* of the minor $\det A[K, L]$. This definition generalizes the definitions of cofactor and algebraic complement given in Section 4.4 for elements (i.e., minors of order 1). The generalization of Theorem 4.2.11, the general *Laplace development of the determinant*, is given in the next theorem. It asserts that the determinant of a matrix A can be evaluated by first choosing a set K of rows and then summing up the products of each minor formed out of those rows with its algebraic complement. A similar development results by replacing rows with columns.

Theorem 4.5.1 *Let $K \subset \{1, 2, \ldots, n\}$ with $|K| = \nu$. Then*

$$\det A = \sum_{L \subseteq \{1,2,\ldots,n\}, |L|=\nu} \det A[K, L]\Delta[K, L], \tag{4.17}$$

where, as indicated, the summation is taken over all the $\binom{n}{\nu}$ subsets L of $\{1, 2, \ldots, n\}$ of cardinality ν.

Proof. We use the formula given in (4.16),

$$\det A = \sum_{F \in \mathcal{F}(A)} (-1)^{q(F)} w(F), \tag{4.18}$$

that evaluates the determinant in terms of the König digraph $G(A)$.

Let $K = \{k_1, k_2, \ldots, k_\nu\}$. A 1-factor F in $\mathcal{F}(A)$ contains one edge leaving each black vertex. Let the edges of F leaving the black vertices with labels in K terminate in those white vertices whose set of labels is $L = \{l_1, l_2, \ldots, l_\nu\}$. We partition the 1-factors in $\mathcal{F}(A)$ into $\binom{n}{\nu}$ sets by putting in $\mathcal{F}_L(A)$ all those 1-factors F with the same L. Thus we may write (4.18) as

$$\det A = \sum_{L \subseteq \{1,2,\ldots,n\}, |L|=\nu} \sum_{F \in \mathcal{F}_L(A)} (-1)^{q(F)} w(F).$$

Each 1-factor F in $\mathcal{F}_L(A)$ consists of a 1-factor F_L in the subgraph of G induced on the black vertices with labels in K and white vertices with labels in L, and a 1-factor $F_{\overline{L}}$ joining the black vertices with labels in the complement $\overline{K}$ and white vertices with labels in the complement $\overline{L}$. We see that F_L corresponds to a 1-factor of $G(A[K, L])$ with the same weight, and $F_{\overline{L}}$ corresponds to a 1-factor of $G(A[\overline{K}, \overline{L}]$ with the same weight. We also see that

$$q(F) = q(F_L) + q(F_{\overline{L}}) + t,$$

where t is the number of intersections of edges in F_L with edges in $F_{\overline{L}}$. If we switch the places of a black vertex in K with a black vertex in $\overline{K}$ immediate to its left, we reduce the number of intersections of edges in F_L with edges in $F_{\overline{l}}$ by 1. Hence, by

$$r = (k_1 - 1) + (k_2 - 2) + \cdots + (k_\nu - \nu)$$

switches, we bring $k_1, k_2, \ldots, k_\nu$ to the first ν positions, reducing the number of intersections by r. Similarly, by

$$s = (l_1 - 1) + (l_2 - 2) + \cdots + (l_\nu - \nu)$$

switches of white vertices, we bring $l_1, l_2, \ldots, l_\nu$ to the first ν positions, reducing the number of intersections by s. We conclude that

$$\begin{aligned} t &= (k_1 - 1) + (k_2 - 2) + \cdots + (k_\nu - \nu) + \\ &\quad (l_1 - 1) + (l_2 - 2) + \cdots + (l_\nu - \nu) \\ &= k_1 + k_2 + \cdots + k_\nu + l_1 + l_2 + \cdots + l_\nu + (\text{an even number}). \end{aligned}$$

Now we obtain that $(-1)^{q(F)} w(F)$ equals

$$(-1)^{q(F_L)} w(F_L) \cdot (-1)^{k_1+k_2+\cdots+k_\nu+l_1+l_2+\cdots+l_\nu} (-1)^{q(F_{\overline{L}})} w(F_{\overline{L}}),$$

from which formula 4.17 now follows. □

Example 4.5.2 Let

$$A = \begin{bmatrix} 1 & 0 & 2 & 0 \\ 3 & 1 & 1 & 0 \\ 0 & 1 & 3 & 2 \\ 2 & 0 & 0 & 1 \end{bmatrix}.$$

In Theorem 4.5.1, let $K = \{1,2\}$. Then

$$\begin{aligned}
\det A &= (-1)^{1+2+1+2}\det\begin{bmatrix}1&0\\3&1\end{bmatrix}\det\begin{bmatrix}3&2\\0&1\end{bmatrix}\\
&\quad + (-1)^{1+2+1+3}\det\begin{bmatrix}1&2\\3&1\end{bmatrix}\det\begin{bmatrix}1&2\\0&1\end{bmatrix}\\
&\quad + (-1)^{1+2+1+4}\det\begin{bmatrix}1&0\\3&0\end{bmatrix}\det\begin{bmatrix}1&3\\0&0\end{bmatrix}\\
&\quad + (-1)^{1+2+2+3}\det\begin{bmatrix}0&2\\1&1\end{bmatrix}\det\begin{bmatrix}0&2\\2&1\end{bmatrix}\\
&\quad + (-1)^{1+2+2+4}\det\begin{bmatrix}0&0\\1&0\end{bmatrix}\det\begin{bmatrix}0&3\\2&0\end{bmatrix}\\
&\quad + (-1)^{1+2+3+4}\det\begin{bmatrix}2&0\\1&0\end{bmatrix}\det\begin{bmatrix}0&1\\2&0\end{bmatrix}\\
&= 3+5+0+8+0-0\\
&= 16.
\end{aligned}$$

□

4.6 Exercises

1. Let $A = [a_{ij}]$ be the matrix of order $2n+1$ such that $a_{ij} = 0$ whenever $i+j$ is an even integer. Prove that $\det A = 0$.

2. Let A be a matrix of order n and let k be a positive integer. Show that if $\det(A^k) = 0$, then $\det A = 0$.

3. Calculate the determinant of the matrix

$$\Delta_n(a_1, a_2, \ldots, a_n) = \begin{bmatrix} a_1 & 1 & 1 & \cdots & 1\\ 1 & a_2 & 1 & \cdots & 1\\ 1 & 1 & a_3 & \cdots & 1\\ \vdots & \vdots & \vdots & \ddots & \vdots\\ 1 & 1 & 1 & \cdots & a_n \end{bmatrix}$$

in which all the off-diagonal entries equal 1.

4. Prove that

$$\det \begin{bmatrix} a_0 & a_1 & a_2 & \cdots & a_{n-1} & a_n \\ -1 & x & 0 & \cdots & 0 & 0 \\ 0 & -1 & x & \cdots & 0 & 0 \\ \vdots & \vdots & \vdots & \ddots & \vdots & \vdots \\ 0 & 0 & 0 & \cdots & -1 & x \end{bmatrix} = \sum_{i=0}^{n} a_i x^{n-i}.$$

5. Let

$$A = \begin{bmatrix} 0 & -1 & 0 \\ 0 & 0 & -1 \\ -3 & -1 & 3 \end{bmatrix}.$$

Find all values of λ for which $\det(A + \lambda I_3) = 0$.

6. Let A be a matrix of order n with a $k \times (n-k+1)$ zero submatrix for some k with $k = 1, 2, \ldots, n-1$. Show that $\det A = 0$.

7. From the fact that the matrix of order $n \geq 2$ with all entries equal to 1 has determinant equal to 0, conclude that the number of odd permutations of $\{1, 2, \ldots, n\}$ equals the number of even permutations.

8. Compute the determinant of the matrix

$$\begin{bmatrix} 1 & 2 & -3 & 1 \\ 2 & 0 & 1 & 0 \\ 3 & 1 & 2 & -1 \\ 0 & 1 & 0 & 3 \end{bmatrix}$$

using the Laplace development along rows $\{2, 4\}$.

9. Compute the determinant of the matrix in the previous exercise using the Laplace development along columns $\{1, 2\}$.

10. Calculate the determinant of the matrix $A^3 B^3 A^2$, where

$$A = \begin{bmatrix} 1 & 0 & 3 \\ 2 & -1 & -2 \\ 1 & 3 & 1 \end{bmatrix} \text{ and } B = \begin{bmatrix} 0 & 3 & 1 \\ 1 & 0 & -3 \\ 2 & 1 & 4 \end{bmatrix}.$$

11. Prove that

$$\det\begin{bmatrix} 1 & x_1 & x_1^2 & \cdots & x_1^{n-1} \\ 1 & x_2 & x_2^2 & \cdots & x_2^{n-1} \\ 1 & x_3 & x_3^2 & \cdots & x_3^{n-1} \\ \vdots & \vdots & \vdots & \ddots & \vdots \\ 1 & x_n & x_n^2 & \cdots & x_n^{n-1} \end{bmatrix} = \prod_{1\le i<j\le n}(x_j - x_i).$$

This determinant is called the *Vandermonde determinant.*

12. A matrix $A = [a_{ij}]$ of order n is *skew-symmetric* provided that $a_{ij} + a_{ji} = 0$ for all $i \neq j$. Thus each entry on the main diagonal of A equals 0. Prove that the determinant of a skew-symmetric matrix of odd order n is 0.

13. Use the Binet–Cauchy formula to evaluate the determinant of

$$\begin{bmatrix} 1 & 0 & 2 & 3 \\ 2 & 1 & 0 & 1 \end{bmatrix}\begin{bmatrix} 2 & 1 \\ 1 & 0 \\ 0 & 3 \\ 0 & 1 \end{bmatrix}.$$

14. Use the Laplace development and the Binet–Cauchy formula to show that if A is an m by n matrix and B is an n by m matrix, then

$$\det\begin{bmatrix} O & A \\ -B & O \end{bmatrix} = \det(AB).$$

15. Let $a \neq b$. Show that

$$\det\begin{bmatrix} a+b & ab & 0 & \cdots & 0 & 0 \\ 1 & a+b & ab & \cdots & 0 & 0 \\ 0 & 1 & a+b & \cdots & 0 & 0 \\ \vdots & \vdots & \vdots & \ddots & \vdots & \vdots \\ 0 & 0 & 0 & \cdots & a+b & ab \\ 0 & 0 & 0 & \cdots & 1 & a+b \end{bmatrix} = \frac{a^{n+1} - b^{n+1}}{a-b}.$$

Chapter 5

Matrix Inverses

In this chapter we define the inverse of a square matrix and study some of its properties. We give a formula for the inverse in terms of determinants, and then give an interpretation in terms of graphs.

Section 5.1 introduces the concept of the adjoint of a square matrix and establishes some of its properties that enable a construction of the inverse in Section 5.2. It is proved that a square matrix has an inverse if and only if the determinant of the matrix is different from zero. In Section 5.3, cofactors of matrix entries are interpreted by special subgraphs of the Coates digraph associated with the matrix, and this finally leads to a graph-theoretical formula for the entries of the inverse.

5.1 Adjoint and Its Determinant

Let $A = [a_{ij}]$ be a matrix of order n. We recall from Chapter 4 that the cofactor α_{ij} of the element a_{ij} is defined by

$$\alpha_{ij} = (-1)^{i+j} \det A_{ij},$$

where A_{ij} is the matrix of order $n-1$ obtained from A by deleting row i and column j. We also recall the developments of the determinant along rows and columns given by

$$\det A = \sum_{j=1}^{n} a_{ij}\alpha_{ij} = \sum_{j=1}^{n} (-1)^{i+j} a_{ij} \det A_{ij} \quad (i = 1, 2, \ldots, n), \tag{5.1}$$

and

$$\det A = \sum_{i=1}^{n} a_{ij}\alpha_{ij} = \sum_{i=1}^{n}(-1)^{i+j} a_{ij} \det A_{ij} \quad (j = 1, 2, \ldots, n). \tag{5.2}$$

Consider the sum

$$\sum_{j=1}^{n} a_{ij}\alpha_{kj}, \text{ where } k \neq i. \tag{5.3}$$

The cofactors α_{kj} occuring in this sum with $k \neq i$ do not depend on row k of A, since row k is deleted in their definitions. Thus we can replace row k of A by any row whatsoever without changing the α_{kj} and (5.3). If we replace row k by row i in A giving a matrix A' in which row i of A appears twice, then no change occurs in (5.3), but now it represents the development of the determinant of A' along its ith row. Because A' has two identical rows, its determinant equals zero. Thus we have

$$0 = \sum_{j=1}^{n} a_{ij}\alpha_{kj} = \sum_{j=1}^{n} a_{ij}(-1)^{k+j} \det A_{kj}, \quad (k \neq i), \tag{5.4}$$

and in a similar way we conclude that

$$0 = \sum_{i=1}^{n} a_{ij}\alpha_{ik} = \sum_{i=1}^{n} a_{ij}(-1)^{i+k} \det A_{ik}, \quad (k \neq j). \tag{5.5}$$

We summarize what we have shown in the next theorem.

Theorem 5.1.1 *Let $A = [a_{ij}]$ be a matrix of order n. Then*

$$\sum_{j=1}^{n} a_{ij}(-1)^{k+j} \det A_{kj} = \begin{cases} \det A & \text{if } k = i, \\ 0 & \text{if } k \neq i. \end{cases}$$

Similarly,

$$\sum_{i=1}^{n} a_{ij}(-1)^{i+k} \det A_{ik} = \begin{cases} \det A & \text{if } k = j, \\ 0 & \text{if } k \neq j. \end{cases}$$

□

In words, Theorem 5.1.1 asserts that the sum of the products of the entries in a row, respectively, column, of a matrix times the cofactors of the entries in a *different* row, respectively, column, equals zero.

We now make a definition that will enable us to write the equations in Theorem 5.1.1 in a more compact matrix form.

Definition 5.1.2 The *adjoint* of the matrix $A = [a_{ij}]$ of order n is the matrix

$$\text{adj } A = \begin{bmatrix} \alpha_{11} & \alpha_{12} & \cdots & \alpha_{1n} \\ \alpha_{21} & \alpha_{22} & \cdots & \alpha_{2n} \\ \vdots & \vdots & \ddots & \vdots \\ \alpha_{n1} & \alpha_{n2} & \cdots & \alpha_{nn} \end{bmatrix}^T$$

obtained by replacing each entry a_{ij} of A by its cofactor $\alpha_{ij} = (-1)^{i+j} \det A_{ij}$ and then transposing the resulting matrix. □

Example 5.1.3 The adjoint of the general matrix

$$A = \begin{bmatrix} a & b \\ c & d \end{bmatrix}$$

of order 2 is given by

$$\text{adj } A = \begin{bmatrix} d & -c \\ -b & a \end{bmatrix}^T = \begin{bmatrix} d & -b \\ -c & a \end{bmatrix}.$$

The adjoint of the matrix

$$A = \begin{bmatrix} 1 & 0 & 2 \\ 2 & 1 & 0 \\ 0 & 2 & 1 \end{bmatrix}$$

is given by

$$\text{adj } A = \begin{bmatrix} 1 & -2 & 4 \\ 4 & 1 & -2 \\ -2 & 4 & 1 \end{bmatrix}^T = \begin{bmatrix} 1 & 4 & -2 \\ -2 & 1 & 4 \\ 4 & -2 & 1 \end{bmatrix}.$$

The adjoint of the identity matrix I_n is I_n itself. This follows since deleting row i and column j with $i \neq j$ always results in a matrix with a zero row (and a zero column) and hence in a matrix whose determinant equals 0, while deleting row and column i results in I_{n-1}, a matrix with determinant equal to 1. □

Let adj $A = [\beta_{ij}]$ so that $\beta_{ij} = \alpha_{ji}$ for each i and j. Then the equations in Theorem 5.1.1 can be written in the following forms:

$$\sum_{j=1}^{n} a_{ij}\beta_{jk} = \sum_{j=1}^{n} a_{ij}(-1)^{k+j} \det A_{kj} = \begin{cases} \det A & \text{if } k = i, \\ 0 & \text{if } k \neq i, \end{cases}$$

and

$$\sum_{i=1}^{n} \beta_{ki}a_{ij} = \sum_{i=1}^{n} a_{ij}(-1)^{i+k} \det A_{ik} = \begin{cases} \det A & \text{if } k = i, \\ 0 & \text{if } k \neq i. \end{cases}$$

These two sets of equations now give us the matrix equation in the next theorem.

Theorem 5.1.4 *If A is a square matrix of order n, then*

$$A(\text{adj } A) = (\text{adj } A)A = (\det A)I_n.$$

□

Example 5.1.5 Continuing with the matrix A of order 3 in Example 5.1.3, we calculate that $\det A = 9$,

$$A(\text{adj } A) = \begin{bmatrix} 1 & 0 & 2 \\ 2 & 1 & 0 \\ 0 & 2 & 1 \end{bmatrix} \begin{bmatrix} 1 & 4 & -2 \\ -2 & 1 & 4 \\ 4 & -2 & 1 \end{bmatrix} = 9I_3,$$

and

$$(\text{adj } A)A = \begin{bmatrix} 1 & 4 & -2 \\ -2 & 1 & 4 \\ 4 & -2 & 1 \end{bmatrix} \begin{bmatrix} 1 & 0 & 2 \\ 2 & 1 & 0 \\ 0 & 2 & 1 \end{bmatrix} = 9I_3.$$

□

5.2 Inverse of a Square Matrix

We begin with the definition of an inverse of a matrix.

Definition 5.2.1 Let $A = [a_{ij}]$ be a square matrix of order n. A matrix $B = [b_{ij}]$ of order n is an *inverse* of A provided $AB = BA = I_n$. An inverse of a matrix A is denoted by A^{-1}. If the matrix A has an inverse, then A is called *invertible* and is also sometimes called *nonsingular*. A *singular matrix* is a square matrix that does not have an inverse.

In order that the notation for the inverse of a matrix not be ambiguous, we need to know that if a matrix has an inverse, then it has only one inverse. This, as well as some elementary properties of inverses, are contained in the next theorem.

Theorem 5.2.2 *Let A be a square matrix of order n. Then:*

(i) *A has at most one inverse.*

(ii) *A has an inverse if and only if* $\det A \neq 0$. *If* $\det A \neq 0$, *then*

$$A^{-1} = \frac{1}{\det A}(\text{adj } A).$$

(iii) *If B is a matrix of order n such that $AB = I_n$, then also $BA = I_n$ and B is the inverse of A.*

Proof. (i) Suppose that both B and C satisfy the definition of an inverse of A. Then we calculate that

$$B = BI_n = B(AC) = (BA)C = I_nC = C.$$

Thus $B = C$ and A has at most one inverse.

(ii) First suppose that A has an inverse B. Then $AB = I_n$. By the multiplicative property of determinants,

$$\det A \det B = \det AB = \det I_n = 1.$$

Hence $\det A \neq 0$. Conversely, suppose that $\det A \neq 0$. Then by Theorem 5.1.4 we have

$$A(\text{adj } A) = (\text{adj } A)A = (\det A)I_n.$$

Because $\det A \neq 0$, we get

$$A\left(\frac{1}{\det A}(\text{adj } A)\right) = \left(\frac{1}{\det A}(\text{adj } A)\right) A = I_n,$$

and hence

$$A^{-1} = \frac{1}{\det A}(\text{adj } A).$$

(iii) Suppose B is a matrix of order n with $AB = I_n$. By the multiplicative property of determinants again,

$$\det A \det B = \det AB = \det I_n = 1.$$

Hence $\det A \neq 0$ and, by (ii), A has an inverse A^{-1}. We calculate that

$$A^{-1} = A^{-1}I_n = A^{-1}(AB) = (A^{-1}A)B = I_nB = B.$$

Hence $B = A^{-1}$. □

Example 5.2.3 Continuing with the matrix

$$A = \begin{bmatrix} 1 & 0 & 2 \\ 2 & 1 & 0 \\ 0 & 2 & 1 \end{bmatrix}$$

in Example 5.1.5, we have that

$$A^{-1} = \frac{1}{9}\begin{bmatrix} 1 & 4 & -2 \\ -2 & 1 & 4 \\ 4 & -2 & 1 \end{bmatrix} = \begin{bmatrix} 1/9 & 4/9 & -2/9 \\ -2/9 & 1/9 & 4/9 \\ 4/9 & -2/9 & 1/9 \end{bmatrix}.$$

□

5.3 Graph-Theoretic Interpretation

In this section we give a formula for the inverse of an invertible matrix in terms of its Coates digraph. In (ii) of Theorem 5.2.2, the elements of the inverse are expressed in terms of the cofactors and the determinant. In Chapter 4, we evaluated the determinant in terms of the Coates digraph, and so we now consider the cofactors from the viewpoint of the Coates digraph.

First we introduce the idea of a 1-connection of a digraph.

Definition 5.3.1 Let D be a digraph with vertices $1, 2, \ldots, n$. Let i and j be vertices of D. A 1-*connection* of vertex i to vertex j is a spanning subdigraph $D[i \to j]$ of D with the following properties:

If $i \neq j$, then

(i) exactly one edge leaves, but no edge enters, vertex i;

(ii) exactly one edge enters, but no edge leaves, vertex j;

(iii) for each vertex $k \neq i, j$, exactly one edge enters, and exactly one edge leaves, vertex k.

If $i = j$, then

(i) no edges enter or leave vertex i;

(ii) for each vertex $k \neq i$, exactly one edge enters, and exactly one edge leaves, vertex k.

□

It follows from the definition that a 1-connection $D[i \to j]$ is a spanning subdigraph of D consisting of a path from i to j (this path is a path of length 0, that is, it is the single vertex i, if $i = j$) and a possibly empty collection of pairwise vertex disjoint cycles having no vertex in common with the path. We let $c(D[i \to j])$ denote the number of cycles of $D[i \to j]$. As usual, if D is a weighted digraph, then the weight $w(D[i \to j])$ of $D(i \to j)$ is the product of the weights of its edges.

In Figure 5.1, a digraph is displayed with several 1-connections of it.

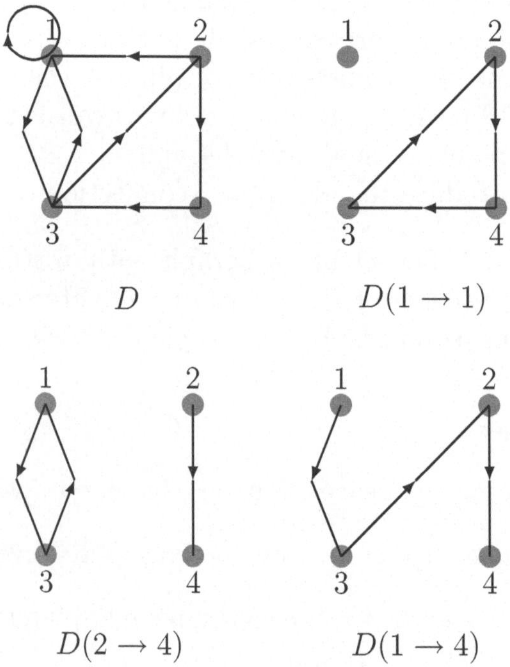

Figure 5.1

Let $A = [a_{ij}]$ be a square matrix of order n and let $D^* = D^*(A)$ be the Coates digraph of A whose edges are weighted by the entries of A. There is a close relationship between the linear subdigraphs of the digraph D^* and their 1-connections. Let L be a linear subdigraph of D^*, and let L contain the edge from vertex j to vertex i of weight a_{ij}. Suppose that we delete from L this edge from vertex j to vertex i. It follows from the definitions of a 1-connection and of a linear subdigraph that the result is a 1-connection $D^*[i \to j]$. If $i = j$, the edge deleted is a loop at vertex i. The following relationships hold between the number of cycles and the weight of the linear subdigraph L and a corresponding 1-connection $D^*[i \to j]$:

$$c(L) = c(D^*[i \to j]) + 1, \tag{5.6}$$

and

$$w(L) = a_{ij} w(D^*[i \to j]). \tag{5.7}$$

Conversely, if $D^*[i \to j]$ is a 1-connection from i to j, then, by adding the edge from vertex j to vertex i, we obtain a linear subdigraph of D satisfying (5.6) and (5.7).

According to Theorem 4.2.11, the cofactor α_{ij} of the element a_{ij} of A is the coefficient of a_{ij} in the development of the determinant of A along row i. Let $\mathcal{L}_{ij}$ denote the set of all linear subdigraphs of D^* containing the edge from j to i. Then, from the definition of a determinant as given in (4.1), we have

$$\alpha_{ij} a_{ij} = (-1)^n \sum_{L \in \mathcal{L}_{ij}(A)} (-1)^{c(L)} w(L), \tag{5.8}$$

where the summation extends over all linear subdigraphs L in $\mathcal{L}_{ij}(A)$. Using (5.6) and (5.7), we get from (5.8) that

$$\alpha_{ij} = (-1)^n \sum_{D^*[i \to j]} (-1)^{c(D^*[i \to j])+1} w(D^*[i \to j]), \tag{5.9}$$

where the summation extends over all 1-connections $D^*[i \to j]$ of D^* from i to j.

We now obtain the following formula for the entries of an invertible matrix.

Theorem 5.3.2 *Let $A = [a_{ij}]$ be an invertible matrix of order n, and let $A^{-1} = [a'_{ji}]$. Then*

$$a'_{ji} = \frac{\sum_{D^*[i \to j]} (-1)^{c(D^*[i \to j])+1} w(D^*[i \to j])}{\sum_{L \in \mathcal{L}(A)} (-1)^{c(L)} w(L)}, \quad (1 \le i, j \le n). \tag{5.10}$$

Proof. By (ii) of Theorem 5.2.2,

$$A^{-1} = \frac{1}{\det A} \text{adj } A,$$

that is,

$$a'_{ji} = \frac{\alpha_{ij}}{\det A}.$$

Substituting for α_{ij} the formula given in (5.9) and for $\det A$ the formula given in (4.1), we get (5.10) by cancellation of $(-1)^n$. □

Example 5.3.3 Let

$$A = \begin{bmatrix} 0 & 1 & 0 & 0 & \cdots & 0 \\ 0 & 0 & 1 & 0 & \cdots & 0 \\ \vdots & \vdots & \vdots & \vdots & \cdots & \vdots \\ 0 & 0 & 0 & 0 & \cdots & 1 \\ -a_n & -a_{n-1} & -a_{n-2} & -a_{n-3} & \cdots & -a_1 \end{bmatrix}$$

whose Coates digraph $D^*(A)$ is given in Figure 5.2, where all horizontal edges from right to left have weight 1.

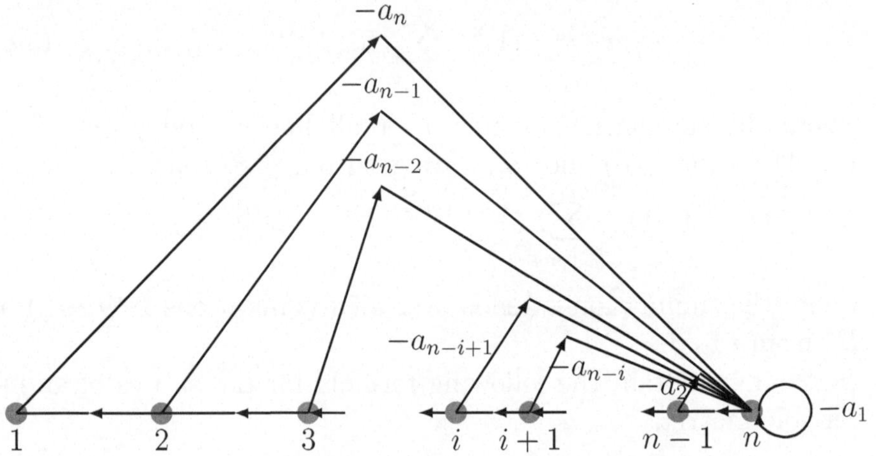

Figure 5.2

The digraph $D^*(A)$ has only one linear subdigraph, and we get $\det A = (-1)^n(-1)^1(-a_n) = (-1)^n a_n$. The only 1-connections of $D^*(A)$ are the 1-connection $D^*(A)[i \to 1]$ with the weight $-a_{n-i}$ for $i = 1, 2, \ldots, n-1$, the 1-connection $D^*(A)[n \to 1]$ with the weight 1, and the 1-connection $D^*(A)[i \to i+1]$ with the weight $-a_n$ for $i = 1, 2, \ldots, n$. Hence

$$A^{-1} = \begin{bmatrix} \frac{a_{n-1}}{a_n} & \frac{-a_{n-2}}{a_n} & \frac{a_{n-3}}{a_n} & \cdots & \frac{a_1}{a_n} & \frac{1}{a_n} \\ 1 & 0 & 0 & \cdots & 0 & 0 \\ 0 & 1 & 0 & \cdots & 0 & 0 \\ \vdots & \vdots & \vdots & \vdots & \vdots & \vdots \\ 0 & 0 & 0 & \cdots & 1 & 0 \end{bmatrix}.$$

□

5.4 Exercises

1. Let A be an invertible matrix of order n and let B be a matrix of order n.

   (a) Prove that $\det A^{-1} = (\det A)^{-1}$.

   (b) Prove that A^T is invertible and $(A^T)^{-1} = (A^{-1})^T$.

   (c) If $\det A = 3$ and $\det B = 4$, what is $\det(A^{-1}BA^3B^2)$?

2. Determine the inverse of a permutation matrix.

3. Prove that a triangular matrix is invertible if and only if it does not have any zeros on its main diagonal.

4. Prove that the inverse of an invertible triangular matrix is triangular.

5. Let A be a matrix of order n. Prove that if A is not invertible, then neither is adj A.

6. Let A be a matrix of order n. Prove that $\det(\text{adj } A) = (\det A)^{n-1}$.

7. Let A and B be invertible matrices of order n. Prove that AB is invertible, indeed, that $(AB)^{-1} = B^{-1}A^{-1}$.

8. Prove that

$$\det \begin{bmatrix} a_{11} & a_{12} & \cdots & a_{1n} & x_1 \\ a_{21} & a_{22} & \cdots & a_{2n} & x_2 \\ \vdots & \vdots & \ddots & \vdots & \vdots \\ a_{n1} & a_{n2} & \cdots & a_{nn} & x_n \\ x_1 & x_2 & \cdots & x_n & 0 \end{bmatrix}$$

   equals

$$-\left(\sum_{i=1}^{n} \alpha_{ii} x_i^2 + \sum_{1 \le i < j \le n} (\alpha_{ij} + \alpha_{ji}) x_i x_j \right).$$

   Here α_{ij} is the cofactor of the element a_{ij} in the matrix $A = [a_{ij}]$ of order n.

9. Find the inverses of each of the following matrices of order n:

   (a) $A = [a_{ij}]$, where $a_{ij} = \alpha_i \delta_{ij}$;

   (b) $B = [b_{ij}]$, where $b_{ij} = \beta_i \delta_{i,n+1-j}$;

   (c) $C = [c_{ij}]$, where $b_{ij} = \gamma_i \delta_{ij} + \delta_{i,j-1}$.

10. Determine the inverse of the matrix

$$A = \begin{bmatrix} c & 1 & 0 & 0 \\ 0 & c & 1 & 0 \\ 0 & 0 & c & 1 \\ 0 & 0 & 0 & c \end{bmatrix}.$$

Chapter 6

Systems of Linear Equations

First, we give a brief introduction to the solution of a system of m linear equations in n unknowns. In particular, we introduce the so-called *reduced row-echelon form* of a matrix and explain how it can be used in solving a system of linear equations. Then, using results from Chapter 5 on the adjoint and inverse of a square matrix, we derive an explicit formula (known as Cramer's formula) for the solution of a linear system of n equations in n unknowns whose coefficient matrix is invertible. We then turn to graph-theoretical techniques for solving systems. In Section 6.3 we show how to use the Coates digraph (flow digraph) to solve the linear system. In the next section, we discuss the signal flow digraph approach (a variation of the previous technique) for solving a linear system. These two techniques, although valid in general, are efficient if the system matrix is *sparse*, that is, if it contains a lot of zero entries and the other entries are variables. Finally, in the last section we explain how to use graph-theoretical tools to treat systems with sparse matrices whose entries are given numerically.

6.1 Solutions of Linear Systems

We begin with some definitions.

Definition 6.1.1 Let F be a field. A linear system of m equations in n unknowns is a system

$$\begin{aligned} a_{11}x_1 + a_{12}x_2 + \cdots + a_{1nx_n} &= b_1 \\ a_{21}x_1 + a_{22}x_2 + \cdots + a_{2n}x_n &= b_2 \\ \vdots \quad &= \vdots \\ a_{m1}x_1 + a_{m2}x_2 + \cdots + a_{mn}x_n &= b_m, \end{aligned}$$

or, in matrix form,

$$Ax = b, \tag{6.1}$$

where

$$A = \begin{bmatrix} a_{11} & a_{12} & \cdots & a_{1n} \\ a_{21} & a_{22} & \cdots & a_{2n} \\ \vdots & \vdots & \ddots & \vdots \\ a_{m1} & a_{m2} & \cdots & a_{mn} \end{bmatrix}$$

is the *matrix of coefficients* and

$$x = \begin{bmatrix} x_1 \\ x_2 \\ \vdots \\ x_n \end{bmatrix} \text{ and } b = \begin{bmatrix} b_1 \\ b_2 \\ \vdots \\ b_m \end{bmatrix}$$

are the matrix-columns of unknowns and constant terms, respectively. The *solution set* of the system (6.1) is the set of all column vectors $x = u$ such that $Au = b$. The system may be *consistent* and have at least one solution, or *inconsistent* and have no solutions. If $b = 0$, then (6.1) is called a *homogeneous system*; otherwise, it is called an *inhomogeneous system.* The homogeneous system $Ax = 0$ is always consistent as $x = 0$ is always a solution; it is for this reason that $x = 0$ is called the *trivial solution* of $Ax = 0$. The solution set of the homogeneous system $Ax = 0$ is called the *null space* of the matrix A. The null space of A is always nonempty as it contains the zero vector. □

The null space of a matrix is a subspace of F^n. This follows since if u and v are in the null space of A, then $Au = 0$ and $Av = 0$ imply that

$$A(cu + dv) = cAu + dAv = c0 + d0 = 0$$

for every choice of constants c and d. In the next theorem we relate the null space of A to the solution set of $Ax = b$.

Theorem 6.1.2 *Consider a linear system $Ax = b$. Let $x = w$ be a particular solution of $Ax = b$, and let U be the null space of A. Then the solution set of $Ax = b$ is the set*

$$w + U = \{w + u : u \in U\}$$

of all vectors obtained by adding to w a vector u in the null space of A.

Proof. Let u be any vector in the null space of A. Then $A(w+u) = Aw + Au = b + 0 = b$, and thus $x = w + u$ is a solution of $Ax = b$. Conversely, let $x = w'$ be any solution of $Ax = b$, and let $u = w' - w$. Then

$$Au = A(w' - w) = Aw' - Aw = b - b = 0,$$

and so u is a vector in the null space of A. Hence $w' = w + u$, so that w' has the required form. □

It follows from Theorem 6.1.2 that by knowing *one* solution of $Ax = b$ and *all* solutions of $Ax = 0$, we can obtain *all* solutions of $Ax = b$.

In addition to the null space of A there are two[1] other subspaces that we associate with A. The first is the *row space* of A consisting of all vectors spanned by the rows of A; the second is the *column space* of A consisting of all vectors spanned by the columns of A. Let the rows of A be $\alpha_1, \alpha_2, \ldots, \alpha_m$. It follows from the definition of a dot product that the null space of A consists of all those vectors $u = [u_1 \ u_2 \ \ldots \ u_n]^T$ such that $\alpha_i^T \cdot u = 0$, equivalently, $\alpha_i u = 0$, for $i = 1, 2, \ldots, m$. Because the row space of A is spanned by $\alpha_1, \alpha_2, \ldots, \alpha_m$, the null space of A consists of all those vectors u such that $\alpha \cdot u = 0$ for all vectors α in the row space of A.[2]

[1]Actually there are three, but only two of them concern us in this brief introduction. The third is the null space of the transpose A^T of A, that is, all vectors u in F^n such that $A^T u = 0$, equivalently, $u^T A = 0$.

[2]We can turn this around and say that the row space of A consists of all those vectors α such that $\alpha^T \cdot u = 0$ for all vectors u in the null space of A.

Definition 6.1.3 Let $A = [a_{ij}]$ be an m by n matrix. The *row rank* of A is the dimension $\mathrm{rr}(A)$ of the row space of A, equivalently, the maximum number of linearly independent rows of A. The *column rank* of A is the dimension $\mathrm{cr}(A)$ of the column space of A, equivalently, the maximum number of linearly independent columns of A. The *nullity* of A is the dimension $\mathrm{n}(A)$ of the null space of A. □

We now show how to find a basis of the row, column, and null spaces of a matrix A.

Definition 6.1.4 Consider the linear system (6.1) of m equations in n unknowns. There are three types of *elementary operations* that can be performed on (6.1) without changing its set of solutions. These are

I. Switch the order of two equations.

II. Multiply both sides of one equation by a nonzero[3] scalar c.

III. Add a multiple c of one equation to a second equation.

Let

$$A' = [A\ b]$$

be the m by $n+1$ *augmented matrix* of (6.1) obtained by affixing to the coefficient matrix A the column vector b as a last column. Then the elementary operations I, II, and III, when applied to A' or A, are called *elementary row operations* (EROs for short) and can be described as follows:

I. Switch the order of two rows.

II. Multiply a row by a nonzero scalar.

III. Add a multiple of one row to a second row.

□

[3]If we were to multiply by zero, we would wipe out the equation, that is, replace it with $0 = 0$.

It is evident that operations I, II, and III do not change the solution set. This is because each of them is reversible. To reverse a type I operation, switch the equations (rows) back. To reverse a type II operation, multiply the same equation (row) by the reciprocal of c. To reverse a type III operation, add the multiple $-c$ of the first equation (row) to the second equation (row).

EROs can be performed using matrix multiplication. Let $I_m(i,j)$ be the matrix obtained by switching row i and row j of the identity matrix I_m of order m, where $1 \le i < j \le m$. Let $I_m(c \cdot i)$ be the matrix obtained by multiplying row i of I_m by the nonzero scalar c, where $1 \le i \le m$. Let $I_m(c \cdot i + j)$ be the matrix obtained from I_m by adding c times row i to row j, where $1 \le i \ne j \le m$ and c is a scalar.[4] The matrices of these three types are called *elementary matrices.*

Example 6.1.5 To illustrate, we have

$$I_4(2,4) = \begin{bmatrix} 1 & 0 & 0 & 0 \\ 0 & 0 & 0 & 1 \\ 0 & 0 & 1 & 0 \\ 0 & 1 & 0 & 0 \end{bmatrix}, \; I_4(5 \cdot 2) = \begin{bmatrix} 1 & 0 & 0 & 0 \\ 0 & 5 & 0 & 0 \\ 0 & 0 & 1 & 0 \\ 0 & 0 & 0 & 1 \end{bmatrix}, \text{ and}$$

$$I_4(3 \cdot 4 + 2) = \begin{bmatrix} 1 & 0 & 0 & 0 \\ 0 & 1 & 0 & 3 \\ 0 & 0 & 1 & 0 \\ 0 & 0 & 0 & 1 \end{bmatrix}.$$

□

The above remark about reversibility of elementary operations (EROs) can be restated in matrix terms as follows:

$$I_m(i,j)^{-1} = I_m(i,j), I_m(c \cdot i)^{-1} = I_m(c^{-1} \cdot i), \text{ and}$$

$$I_m(c \cdot i + j)^{-1} = I_m(-c \cdot i + j).$$

In particular, the inverse of an elementary matrix is an elementary matrix of the same type.

---

[4]It's acceptable that $c = 0$, but then $I_m(0 \cdot i + j) = I_m$.

Using EROs, a system of linear equations (equivalently, its augmented matrix) can be reduced to a simple form from which the solution set is then evident. The EROs of type III are used to eliminate variables from equations, that is, make the coefficients equal to zero.

Definition 6.1.6 Let R be an m by n matrix. Then R has a *reduced row-echelon form*, abbreviated *rre-form*, provided each of the following properties hold:

(i) Zero rows, if present, come last.

(ii) The first nonzero entry in each nonzero row is a 1, called a *pivot*, and every other entry in the column of that 1 equals 0.

(iii) If there are k nonzero rows and the pivot 1 in row i is in column p_i, then $1 \leq p_1 < p_2 < \cdots < p_k \leq n$. (The matrix R contains the identity matrix I_k as a submatrix.)

□

Example 6.1.7 A zero matrix O is already in rref-form. The matrix

$$\begin{bmatrix} 1 & 3 & 0 & 0 & 4 & 0 \\ 0 & 0 & 1 & 0 & 2 & 0 \\ 0 & 0 & 0 & 1 & 3 & 0 \\ 0 & 0 & 0 & 0 & 0 & 1 \\ 0 & 0 & 0 & 0 & 0 & 0 \\ 0 & 0 & 0 & 0 & 0 & 0 \end{bmatrix}$$

is in rre-form. □

In the next theorem we show that every matrix may be put in rre-form using EROs. This process is often referred to as *Gaussian elimination.*

Theorem 6.1.8 *Let $A = [a_{ij}]$ be an m by n matrix. Then there exists a sequence $P_1, P_2, \ldots, P_s$ of elementary matrices such that $P_s \cdots P_2 P_1 A$ is a matrix in reduced row-echelon form. The matrix $P = P_s \cdots P_2 P_1$, being a product of invertible matrices, is an invertible matrix.*

Proof. We briefly describe the constructive proof in terms of EROs. Let the first nonzero column of A be column p_1. Using a type I ERO, if necessary, we may bring a nonzero entry in column p_1 to row 1. Using a type II ERO, if necessary, we may make that nonzero entry 1. Using type III EROs as necessary, we obtain a matrix with all other entries in column p_1 equal to 0. We now repeat with the submatrix determined by rows $2, \ldots, m$ and columns $p_1 + 1, \ldots, n$, obtaining a matrix with a 1 in row 2 and column $p_2 > p_1$ with all of the other entries in column 2 equal to 0. Using a type III ERO we can also make the entry in row 1 and column p_2 equal to zero. We then consider the submatrix formed by rows $3, \ldots, m$ and columns $p_2+1, \ldots, n$ and continue until only zero rows remain. □

Example 6.1.9 When we obtain the rre-form of the augmented matrix $[A\ b]$ of a system of linear equation, we can immediately read off its set of solutions, or conclude that the system is inconsistent. The system will be inconsistent exactly when one of the pivots occurs in the last column, the column corresponding to b. In this case, one of the equations becomes the contradictory equation $0 = 1$. In case of a homogeneous system, which is always consistent, we can use the coefficient matrix A itself rather than the augmented matrix $[A\ 0]$.

For instance, suppose the rre-form of the augmented matrix of a system $Ax = b$ of linear equations in unknowns $x_1, x_2, \ldots, x_6$ is

$$\left[\begin{array}{cccccc|c} 1 & 3 & 0 & 0 & 4 & 0 & 2 \\ 0 & 0 & 1 & 0 & 2 & 0 & 1 \\ 0 & 0 & 0 & 1 & 3 & 0 & 5 \\ 0 & 0 & 0 & 0 & 0 & 1 & 3 \\ 0 & 0 & 0 & 0 & 0 & 0 & 0 \\ 0 & 0 & 0 & 0 & 0 & 0 & 0 \end{array}\right],$$

where we have drawn a vertical line to separate the last column corresponding to b from the other columns. Thus, with elementary operations, the original system of equations has been reduced to the following system, with the same set of solutions:

$$x_1 + 3x_2 + 4x_5 \;=\; 2$$

$$\begin{aligned} x_3 + 3x_5 &= 5 \\ x_4 + 3x_5 &= 1 \\ x_6 &= 3. \end{aligned}$$

The variables corresponding to the pivots are x_1, x_3, x_4, x_6 and can be solved in terms of the *free* variables x_2, x_5 as follows:

$$\begin{aligned} x_1 &= 2 - 3x_2 - 4x_5 \\ x_3 &= 5 - 3x_5 \\ x_4 &= 5 - 3x_5 \\ x_6 &= 3. \end{aligned} \tag{6.2}$$

Thus x_2 and x_5 can take any values with $x_1, x_3, x_4,$ and x_6 determined by (6.2). The dimension of the null space is the number 2 of free variables. □

The reduction to rre-form gives rise to some important consequences, which we now elaborate on.

Let A be an invertible matrix of order n. Let R be the rre-form of A so that, by Theorem 6.1.8, there is a product of elementary, and so invertible, matrices, $P = P_s \cdots P_2 P_1$ such that $PA = R$. Thus R, being a product of invertible matrices, is also invertible. Hence R cannot have any zero rows. Since A is a square matrix, this means that $R = I_n$, and now $PA = I_n$ implies that $A^{-1} = P_s \cdots P_2 P_1$. Thus the inverse of an invertible matrix can be found by applying EROs to reduce A to I_n. One way to do this is to apply these EROs to the matrix

$$\left[\begin{array}{c|c} A & I_n \end{array}\right].$$

The result is

$$\left[\begin{array}{c|c} I_n & P \end{array}\right] = \left[\begin{array}{c|c} I_n & A^{-1} \end{array}\right].$$

(If the rre-form of A does not equal I_n, then A is not invertible.) From the above discussion, we now obtain the following corollary.

Corollary 6.1.10 *A square matrix of order n is invertible if and only if it is a product of elementary matrices.* □

The row space of an m by n matrix A consisting as it does of all the linear combinations of the rows of A is the set of all vectors of the form $u^T A$ as u ranges over all m by 1 vectors. Let P be an invertible matrix of order m. Then

$$u^T(PA) = (u^T P)A = v^T A,$$

where $v^T = u^T P$; conversely,

$$v^T A = v^T(P^{-1}P)A = (v^T P^{-1})(PA) = u^T(PA),$$

where $u^T = v^T P^{-1}$. These two equations imply that the row space of A is the same as the row space of PA for any invertible matrix P, and hence the row rank of A equals the row rank of PA. In particular, the row rank of A equals the row rank of its rre-form, and this is easily seen to be the number of pivots (number of nonzero rows). In general, the column space does change, but the linear dependence or linear independence of the columns does not. Another way of saying the same thing is that the null space of A equals the null space of PA for every invertible matrix:

If $Au = 0$, so does $(PA)u = 0$; conversely, if $(PA)u = 0$, then multiplying by P^{-1} we see that $Au = 0$.

From this we conclude that the column rank of A equals the column rank of PA. In particular, the column rank of A equals the column rank of its rre-form, and this is also easily seen to be the number of pivots. We conclude that *the row rank and column rank of a matrix are always equal and the common value is the number of pivots in its rref-form.* This common value is called the *rank* of A and is denoted as $\mathrm{r}(A)$. The dimension of the null space is the number of free variables, and this equals $n - \mathrm{r}(A)$.

Suppose the m by n matrix A has an invertible submatrix of order k. Then the k rows (and the k columns) of A containing this submatrix are linearly independent, and so the rank of A is at least equal to k. Conversely, suppose the rank of A equals k. Then A has k linearly independent rows and these form a k by n

submatrix A' of A with rank equal to k. Since the row rank of A' equals its column rank, A' has k linearly independent columns forming a square submatrix B of order k of A' and hence of A. The matrix B has linearly independent rows and so its rre-form is I_k. Hence B is invertible and $\det B \neq 0$. It follows that the rank of A equals the largest order of a submatrix of A that is invertible.

In the next theorem we collect some of these observations.

Theorem 6.1.11 *Let A be an m by n matrix.*

(i) *The row and column ranks of A are equal. The common value is the rank of A, denoted $\mathrm{r}(A)$.*

(ii) *The rank of A plus the nullity of A equals n:*

$$\mathrm{r}(A) + \mathrm{n}(A) = n.$$

(iii) *The rank of A equals the largest integer k such that A has a submatrix of order k whose determinant is not zero, equivalently, the largest order of an invertible submatrix of A.* □

6.2 Cramer's Formula

Consider a linear system of n equations in n unknowns

$$\begin{aligned} a_{11}x_1 + a_{12}x_2 + \cdots + a_{1nx_n} &= b_1 \\ a_{21}x_1 + a_{22}x_2 + \cdots + a_{2n}x_n &= b_2 \\ \vdots \qquad\qquad & \qquad \vdots \\ a_{n1}x_1 + a_{n2}x_2 + \cdots + a_{nn}x_n &= b_n, \end{aligned}$$

or in matrix form

$$Ax = b, \tag{6.3}$$

where

$$A = \begin{bmatrix} a_{11} & a_{12} & \cdots & a_{1n} \\ a_{21} & a_{22} & \cdots & a_{2n} \\ \vdots & \vdots & \ddots & \vdots \\ a_{n1} & a_{n2} & \cdots & a_{nn} \end{bmatrix}$$

is the matrix of coefficients and

$$x = \begin{bmatrix} x_1 \\ x_2 \\ \vdots \\ x_n \end{bmatrix} \text{ and } b = \begin{bmatrix} b_1 \\ b_2 \\ \vdots \\ b_n \end{bmatrix}$$

are the column vectors of unknowns and constant terms, respectively.

We now assume that the coefficient matrix A is invertible. From Chapter 5 we know that

$$A^{-1} = \frac{1}{\det A}(\text{adj } A) = \left[\frac{1}{\det A}\beta_{ij}\right],$$

where

$$\frac{1}{\det A}\beta_{ij} = (-1)^{j+i}\frac{\det A_{ji}}{\det A} \quad (i, j = 1, 2, \ldots, n) \qquad (6.4)$$

and A_{ji} is the submatrix of A of order $n-1$ obtained by deleting row j and column i. Multiplying both sides of (6.3) by A^{-1}, we get

$$A^{-1}(Ax) = A^{-1}b.$$

Since $A^{-1}A = I_n$ and $I_n x = x$, we get that

$$x = A^{-1}b$$

is the unique solution of (6.3). To find the solution we multiply b on its left by A^{-1}. Using the formula for the entries of A^{-1} as given in (6.4), we obtain

$$x_i = \sum_{j=1}^{n}(-1)^{j+i}b_j\frac{\det A_{ji}}{\det A} = \frac{1}{\det A}\sum_{j=1}^{n}(-1)^{j+i}b_j \det A_{ji} \qquad (6.5)$$

for $i = 1, 2, \ldots, n$.

Let $A^{(i)}$ be the matrix of order n obtained from A by replacing its ith column with the column vector b. As discussed in Chapter 4, the cofactors of the entries in column i of $A^{(i)}$ do not depend on what is actually contained in column i. This implies that these cofactors are the same as the corresponding cofactors of the elements in column i of A. It follows that by developing the determinant of $A^{(i)}$ along column i, we get that

$$\det A^{(i)} = \sum_{j=1}^{n}(-1)^{j+i} b_j \det A_{ji} \quad (i = 1, 2, \ldots, n). \tag{6.6}$$

Comparing (6.6) with (6.5), we see that

$$x_i = \frac{\det A^{(i)}}{\det A} \quad (i = 1, 2, \ldots, n). \tag{6.7}$$

This is *Cramer's formula,* which we summarize in the next theorem. It expresses the solution of $Ax = b$ as the quotient of two determinants.

Theorem 6.2.1 *Let $Ax = b$ be a system of linear equations in n unknowns where the matrix A of coefficients is invertible. Let $A^{(i)}$ be the matrix of order n obtained from A by replacing its ith column with the column vector b, $(1 \leq i \leq n)$. Then $Ax = b$ has a unique solution $x = (x_1, x_2, \ldots, x_n)^T$ given by*

$$x_i = \frac{\det A^{(i)}}{\det A}, \quad (i = 1, 2, \ldots, n).$$

□

Example 6.2.2 We solve the system of linear equations

$$\begin{aligned} x_1 + 3x_2 - 4x_3 &= 2 \\ 0x_1 - x_2 + 3x_3 &= 0 \\ 3x_1 + x_2 + x_3 &= -1. \end{aligned}$$

The matrix of coefficients is

$$A = \begin{bmatrix} 1 & 3 & -4 \\ 0 & -1 & 3 \\ 3 & 1 & 1 \end{bmatrix}$$

with determinant equal to 11. Thus A is invertible and the unique solution $x = (x_1, x_2, x_3)^T$ is given by

$$x_1 = (1/11)\det\begin{bmatrix} 2 & 3 & -4 \\ 0 & -1 & 3 \\ -1 & 1 & 1 \end{bmatrix} = \frac{-13}{11},$$

$$x_2 = (1/11)\det\begin{bmatrix} 1 & 2 & -4 \\ 0 & 0 & 3 \\ 3 & -1 & 1 \end{bmatrix} = \frac{21}{11},$$

$$x_3 = (1/11)\det\begin{bmatrix} 1 & 3 & 2 \\ 0 & -1 & 0 \\ 3 & 1 & -1 \end{bmatrix} = \frac{7}{11}.$$

□

6.3 Solving Linear Systems by Digraphs

Electrical engineers have developed a series of methods for solving systems of linear algebraic equations that appear in the theory of electrical circuits, control theory, and other areas. We start in this section by explaining the flow graph method of Coates, known since the 1950s. In the next section we will desccribe the signal flow graph technique of Mason.

As in the previous section, we again consider a system

$$Ax = b, \text{ equivalently } -b + Ax = 0 \tag{6.8}$$

of n linear equations in n unknowns written as one matrix equation. The matrix $A = [a_{ij}]$ is a square matrix of order n.

Definition 6.3.1 The *Coates digraph (also called the flow digraph or simply flow graph) of the linear system* (6.8) is the Coates digraph $D^*(-b, A)$ of the matrix $[-b\ A]$ with $n + 1$ vertices labeled $0, 1, 2, \ldots, n$ whose directed edges are those given by the following rules:

(i) For each i, j with $1 \leq i, j \leq n$ and $a_{ij} \neq 0$, there is an edge from vertex j to vertex i of weight a_{ij}.

(ii) For each i with $1 \leq i \leq n$ and $b_i \neq 0$, there is an edge from vertex 0 to vertex i of weight $-b_i$.

It is to be observed that the subdigraph of $D^*(-b, A)$ induced on the set of vertices $\{1, 2, \ldots, n\}$ is just the Coates digraph $D^*(A)$ of the coefficient matrix A. It is also to be observed that there are no directed edges entering vertex 0. $\square$

Writing, as we have, the system $Ax = b$ in the equivalent way $-b + Ax = 0$ and forming the n by $n+1$ augmented matrix as

$$[-b\ A] = \begin{bmatrix} -b_1 & a_{11} & a_{12} & \cdots & a_{n1} \\ -b_2 & a_{21} & a_{22} & \cdots & a_{n2} \\ \vdots & \vdots & \vdots & \ddots & \vdots \\ -b_n & a_{n1} & a_{n2} & \cdots & a_{nn} \end{bmatrix}, \tag{6.9}$$

we see that we could regard the Coates digraph of $Ax = b$ as being constructed from the matrix $[-b\ A]$, where the vertex 0 corresponds to the initial column. To make this even more precise, we could imagine that an initial row of all zeros has been attached to (6.9) to obtain a square matrix of order $n+1$; the Coates digraph of the linear system $Ax = b$ then becomes the Coates digraph of the resulting matrix of order $n+1$, with vertices labeled $0, 1, 2, \ldots, n$.

Example 6.3.2 The Coates digraph $D^*(-b, A)$ of the system of linear equations

$$\begin{aligned} a_{11}x_1 + a_{12}x_2 + a_{13}x_3 + 0x_4 &= b_1 \\ 0x_1 + a_{22}x_2 + 0x_3 + a_{24}x_4 &= 0 \\ a_{31}x_1 + 0x_2 + a_{33}x_3 + 0x_4 &= b_3 \\ 0x_1 + a_{42}x_2 + a_{43}x_3 + a_{44}x_4 &= 0 \end{aligned} \tag{6.10}$$

is displayed in Figure 6.1, where as usual edges of weight 0 are not shown.

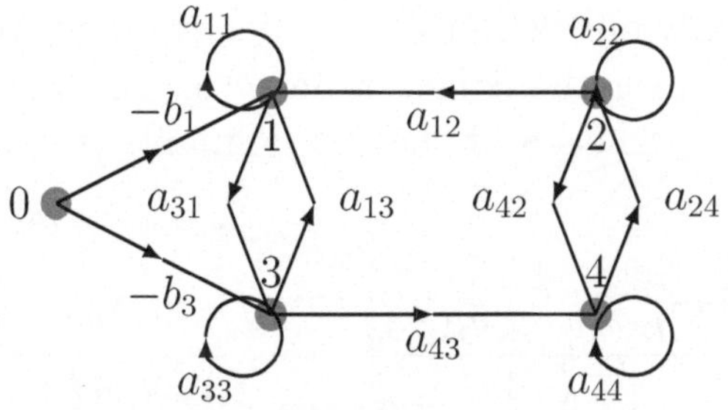

Figure 6.1

The part of the Coates digraph obtained by deleting from $D^*(-b, A)$ the vertex 0 and all the edges that leave vertex 0 is the Coates digraph $D^*(A)$ corresponding to the matrix A. $\square$

We now relate the 1-connections of $D^*(-b, A)$ to those of $D^*(A)$. Let $D^* = D^*(A)$. Let $F = D^*[i \to j]$ be a 1-connection of D^* from i to j. If $b_i \neq 0$, then $D^*(-b, A)$ contains the edge from vertex 0 to vertex i of weight $-b_i$. Appending this edge to $D^*[i \to j]$, we obtain a 1-connection F' of $D^*(-b, A)$ from 0 to j. Conversely, a 1-connection F' of $D^*(-b, A)$ from vertex 0 to vertex j gives, upon deletion of vertex 0 and the unique edge leaving it, a 1-connection F of D^* from some vertex i to vertex j. Because 1-connections of $D^*(-b, A)$ can only go from vertex 0 to some vertex $j \geq 1$, we have a one-to-one correspondence between the 1-connections of $D^*(A)$ and those of $D^*(-b, A)$. The weights of these two 1-connections F and F' under this one-to-one correspondence are related by the formula

$$w(F') = -b_i w(F). \tag{6.11}$$

We now show how to express the solution of (6.3) in terms of its Coates digraph. The key to this is Cramer's formula and the determinant formula given in its definition.

Assume as before that A is invertible, that is, $\det A \neq 0$. Then $Ax = b$ has a unique solution and, by Cramer's formula, this solution is

$$x_i = \frac{\det A^{(i)}}{\det A} = \frac{1}{\det A} \sum_{j=1}^{n} (-1)^{j+i} b_j \det A_{ji} \quad (i = 1, 2, \ldots, n). \tag{6.12}$$

We now apply our definition of a determinant from Section 4.1 and our formula for the cofactors in terms of the Coates digraph as given in (5.9) to each of the determinants given in (6.12). We then obtain

$$x_i = \frac{(-1)^n \sum_{D^*(A)[i\to j]} b_j(-1)^{c(D^*(A)[j\to i])+1} w(D^*(A)[j \to i])}{(-1)^n \sum_{L\in\mathcal{L}(A)} (-1)^{c(L)} w(L)} \tag{6.13}$$

for $i = 1, 2, \ldots, n$, where the summation in the numerator extends over all of the 1-connections $D^*(A)[j \to i]$ of $D^*(A)$ from j to i, and the summation in the denominator extends over all linear subdigraphs L of $D^*(A)$. From our discussion comparing the 1-connections of $D^*(A)$ to those of $D^*(-b, A)$, we can rewrite (6.13) as

$$x_i = \frac{\sum_{D^*(-b,A)[0\to i]} (-1)^{c(D^*(-b,A)[0\to i])} w(D^*(-b\ A)[0 \to i])}{\sum_{L\in\mathcal{L}(A)} (-1)^{c(L)} w(L)} \tag{6.14}$$

for $i = 1, 2, \ldots, n$.

Formula (6.14) is the *Coates formula* for solving a system of n linear equations in n unknowns with an invertible coefficient matrix.

Example 6.3.3 We continue with Example 6.3.2. Figure 6.2 displays all linear subgraphs of the digraph corresponding to the matrix of the system of equations (6.10), while Figure 6.3 displays all 1-connections from the vertex 0 to the vertex 3 of the Coates digraph in Figure 6.1.

Using the formula, we get that the value of x_3 is

$$\frac{b_3a_{11}a_{22}a_{44} - b_3a_{11}a_{42}a_{24} - b_1a_{31}a_{22}a_{44} + b_1a_{31}a_{24}a_{42}}{a_{11}a_{22}a_{33}a_{44} - a_{12}a_{24}a_{43}a_{31} + a_{13}a_{31}a_{42}a_{24} - a_{11}a_{33}a_{42}a_{24} - a_{22}a_{44}a_{13}a_{31}}.$$

□

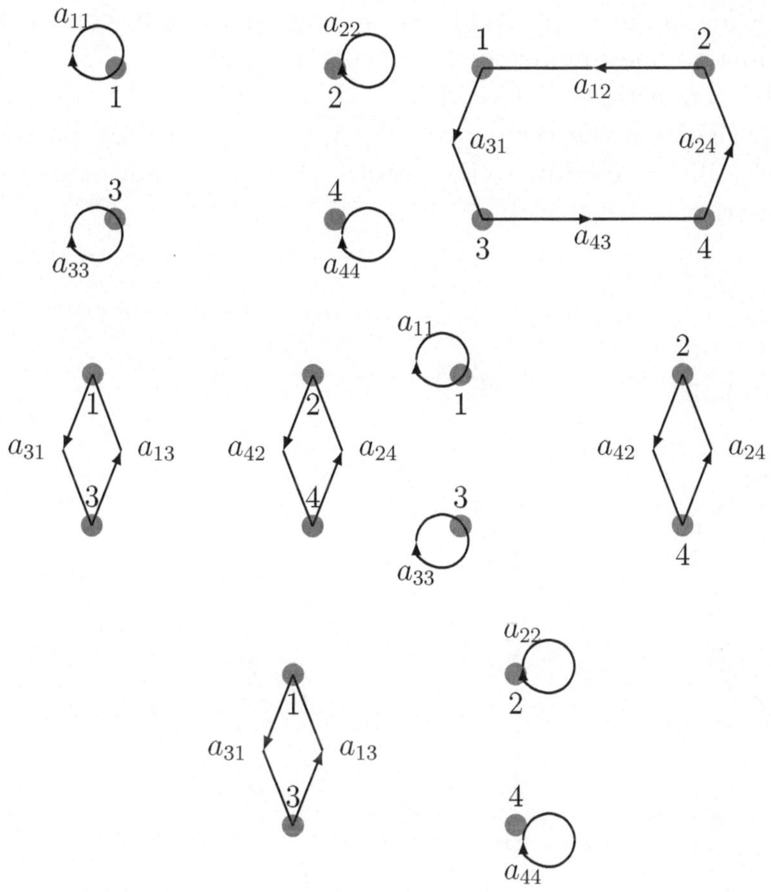

Figure 6.2

The potential advantage of using the Coates formula for solving a system of linear equations results from the fact that the digraph from which we find the solution can be drawn if we are acquainted with the structure of the system described by linear equations, without the need to write the equations; this happens, e.g., in electric circuit theory and control theory. In practice we usually do not list the linear subdigraphs and 1-connections, but determine directly from the Coates digraph the unknown value we are seeking. This requires some experience, because without close attention, some 1-connections or linear subgraphs may be overlooked. Although no efficient general rule for systematically finding the linear subdigraphs and 1-connections is known, the following might help. The linear subdigraphs can be classified according to the number of loops contained in them. In order to find 1-connections $D^*(-b, A)[0 \to i]$ it is necessary to determine all paths from the vertex 0 to the vertex i, and these paths can be classified according to the vertices that come immediately after the vertex 0.

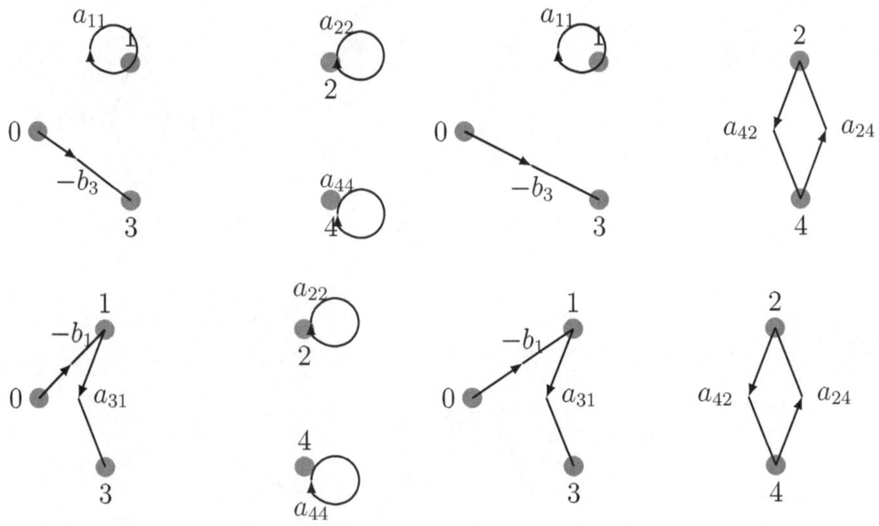

Figure 6.3

The method we have described is intended for calculation by hand. The use of a computer with this method is not recommended, because the power of a computer can be better exploited with other methods that are not suitable for hand calculations. The practical usefulness of this method is limited to systems of equations with not more than about ten unknowns, with the condition that the corresponding digraph has a comparatively small number of edges. Otherwise, it is not practical to find all linear subgraphs and necessary 1-connections in a digraph.

The use of digraphs in solving a system of linear algebraic equations is particularly convenient if the coefficients of the system are not numerical values, as in Example 6.3.3.

6.4 Signal Flow Digraphs of Linear Systems

In this section we discuss a method of Mason for solving certain systems of n linear equations in $n+1$ unknowns $x_0, x_1, x_2, \ldots, x_n$, where x_0 is distinguished as a parameter, with the resulting solution expressed in terms of x_0.

The system of linear equations is assumed to be of the form

$$\begin{aligned}
x_1 &= a_{10}x_0 + a_{11}x_1 + \cdots + a_{1n}x_n,\\
x_2 &= a_{20}x_0 + a_{21}x_1 + \cdots + a_{2n}x_n\\
\vdots & \quad \vdots \\
x_n &= a_{n0}x_0 + a_{n1}x_1 + \cdots + a_{nn}x_n.
\end{aligned} \tag{6.15}$$

The coefficient matrix

$$B = \begin{bmatrix} a_{10} & a_{11} & a_{12} & \cdots & a_{1n} \\ a_{20} & a_{21} & a_{22} & \cdots & a_{2n} \\ \vdots & \vdots & \vdots & \ddots & \vdots \\ a_{n0} & a_{n1} & a_{n2} & \cdots & a_{nn} \end{bmatrix}$$

is an n by $n+1$ matrix. We will also want to consider the initial column

$$A_0 = \begin{bmatrix} a_{10} \\ a_{20} \\ \vdots \\ a_{n0} \end{bmatrix},$$

of B, and the square matrix

$$A = \begin{bmatrix} a_{11} & a_{12} & \cdots & a_{1n} \\ a_{21} & a_{22} & \cdots & a_{2n} \\ \vdots & \vdots & \ddots & \vdots \\ a_{n1} & a_{n2} & \cdots & a_{nn} \end{bmatrix}$$

of order n obtained from B by deleting its initial column.

As with the augmented matrix of a linear system as discussed in Section 6.2, we can imagine that the matrix B has been enlarged by adding an initial row of all zeros, thereby obtaining a square matrix of order $n+1$. Then we consider the Coates digraph $D^*(B)$ with $n+1$ vertices denoted by $x_0, x_1, x_2, \ldots, x_n$ with an edge from vertex x_j to vertex x_i if and only if $b_{ij} \neq 0$. In $D^*(B)$, there are no edges that enter vertex x_0 (as vertex x_0 corresponds to the initial row of zeros we imagined). The digraph $D^*(B)$ obtained in this way is called the *signal flow digraph*, or *Mason's digraph*, of the system (6.15).

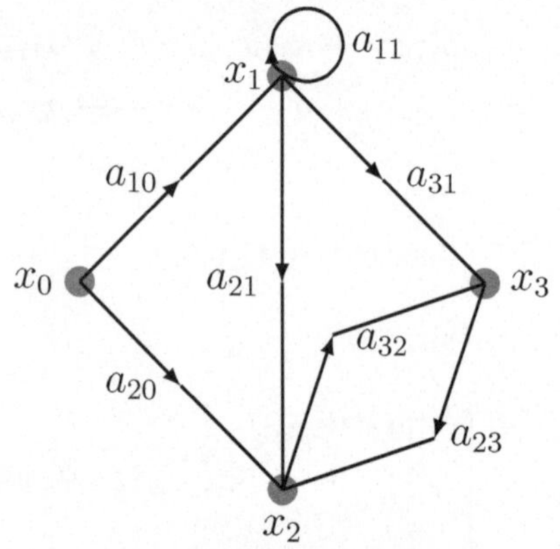

Figure 6.4

Example 6.4.1 The signal flow digraph corresponding to the system of equations

$$\begin{aligned} x_1 &= a_{10}x_0 + a_{11}x_1 \\ x_2 &= a_{20}x_0 + a_{21}x_1 + a_{23}x_3, \\ x_3 &= a_{31}x_1 + a_{32}x_2, \end{aligned}$$

is given in Figure 6.4. □

As usual, we define the weight of paths and cycles to be the product of the weight of their edges.

Definition 6.4.2 Let the directed cycles of the weighted digraph $D^*(A)$ be enumerated by $C_1, C_2, \ldots, C_j, \ldots$ with weights, respectively, $t_1, t_2, \ldots, t_j, \ldots$. Let the paths from vertex x_0 to vertex x_i

be enumerated by $P_1^i, P_2^i, \ldots, P_j^i, \ldots$ $(i = 1, 2, \ldots, n)$. We define $w_j^{(i)}$ to be the weight of P_j^i $(i = 1, 2, \ldots, n; j \geq 1)$.

Mason's determinant $\Delta_M = \Delta_M(D^*(B))$ of the weighted digraph $D^*(B)$ is defined by

$$\Delta_M = 1 - \sum_i{}' t_i + \sum_{i,j}{}' t_i t_j - \sum_{i,j,k}{}' t_i t_j t_k + \cdots, \tag{6.16}$$

where $\sum_i' t_i$ is the sum of the weights of all cycles in the digraph, $\sum_{i,j}' t_i t_j$ is the sum of the products of the weights of all pairs of cycles having no common vertex, $\sum_{i,j,k}' t_i t_j t_k$ is the sum of the products of the weights of all triples of nontouching cycles, that is, cycles no two of which have a common vertex, and so forth. □

Definition 6.4.3 Let G be a weighted digraph with n vertices. Then the *Coates determinant* $\Delta_C = \Delta_C(G)$ of G is defined by

$$\Delta_C = \sum_L (-1)^{c(L)} w(L), \tag{6.17}$$

where the summation extends over all linear subdigraphs of the digraph G. Note that in the case that G is the Coates digraph $D^*(A)$ of a matrix A of order n, then $\Delta_C(D^*(A)) = (-1)^n \det A$.
□

The system of equations (6.15) can be written in the form

$$x = x_0 A_0 + Ax, \text{ equivalently } (A - I_n)x = -x_0 A_0. \tag{6.18}$$

Regarding x_0 as fixed, the augmented matrix of (6.18) is the n by $n + 1$ matrix

$$[x_0 A_0 \; A - I_n] .$$

Thus the Coates digraph $D^*(x_0 A_0, A)$ of the system (6.18) is just the Mason digraph with the weights changed as follows: the weight of each edge leaving the vertex labeled x_0 is obtained by multipling by x_0, and the weight of the loops at vertices labeled $x_1, x_2, \ldots, x_n$ is obtained by subtracting 1. Note that if there wasn't a loop at one of the vertices x_i, $(1 \leq i \leq n)$ (that is, $a_{ii} = 0$), then a loop of weight -1 is inserted; if there was a loop of weight 1 (that is, $a_{ii} = 1$), then the loop disappears.

Our goal is to show that the solution of the system of linear equations (6.15) is given by the following formula, known as *Mason's formula:*

$$x_i = \frac{\sum_j w_j^{(i)} \Delta_j^{(i)}}{\Delta_M} x_0 \quad (i = 1, \ldots, n), \tag{6.19}$$

where the summation extends over all the paths from x_0 to x_i in $D^*(B)$ as enumerated in Definition 6.4.2. In this formula,

(i) $\Delta_j^{(i)}$ denotes Mason's determinant of the subdigraph of $D^*(B)$ obtained by deleting all vertices of the path P_j^i.

(ii) Δ_M is the (ordinary) determinant of the matrix $A - I_n$ of order n, where A, as previously explained, is obtained from B by deleting its initial column.

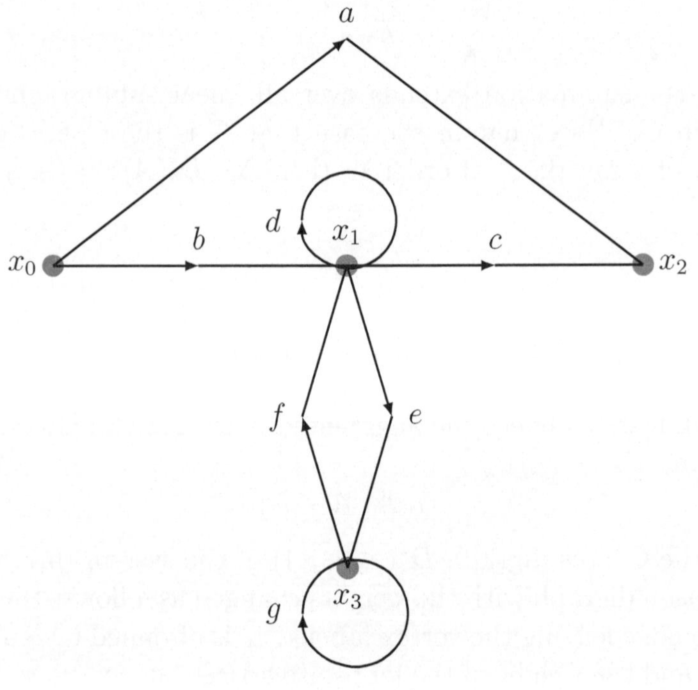

Figure 6.5

Example 6.4.4 Consider the system of linear equations

$$\begin{aligned} x_1 &= bx_0 + dx_1 + fx_3 \\ x_2 &= ax_0 + cx_1 \\ x_3 &= \qquad ex_1 + gx_3, \end{aligned}$$

whose corresponding signal flow digraph is displayed in Figure 6.5.

Using formula (6.19) we get

$$x_2 = \frac{a(1-(d+ef+g)+gd)+bc(1-g)}{1-(d+ef+g)+gd}x_0.$$

□

The introduction of the distinguished variable x_0 is not mathematically necessary but it is useful in applications of signal flow graphs to control theory (see Section 10.1).

Not surprisingly, Mason's formula is derived with the help of the Coates formula.

Using Mason's formula as a model, we can rewrite the Coates formula (6.14) for the solution of a linear system of n equations in n unknowns as

$$x_i = \frac{\sum_k q_k^{(i)} \Delta_C^{(i)}(D^*(-b,A)_k}{\Delta_C(D^*(A))}, \tag{6.20}$$

where $q_k^{(i)}$ denotes the weight of the kth path P_k^i from the vertex 0 to the vertex i and $D^*(-b,A)_k$ is the subdigraph of the Coates digraph obtained by deleting the vertices of this kth path.

Consider the digraph $D^*(A-I_n)$ that arises from the digraph $D^*(A)$ by subtracting 1 from the weight of every loop of $D^*(A)$; as before, if there is no loop at a vertex, then the vertex gets a loop with the weight -1. We first verify the formula

$$\Delta_C(D^*(A-I_n)) = \Delta_M(D^*(A)). \tag{6.21}$$

To see this, first note that

$$\Delta_C(D^*(A-I_n)) = (-1)^n \det(A-I_n).$$

Setting $\lambda = 1$ in the formula given in Corollary 4.3.2, we get that

$$(-1)^n \det(A - I_n) = (-1)^n \left(1 + \sum_{p=1}^{n} (-1)^p \sum_{A_p} \det A_p\right), \quad (6.22)$$

where the second sum is taken over all principal submatrices A_p of A of order p. Now, applying the definition of the determinant to the quantities $\det A_p$, we see that (6.22) gives $\Delta_M(D^*(A))$ as defined in (6.16).

Now consider the system of equations (6.15) written in the form

$$(A - I_n)x = -x_0 A_0$$

as given in (6.18). Solving this system using (6.20), and using (6.21), we get

$$\begin{aligned} x_j &= \frac{\sum_k q_k^{(j)} \Delta_C^{(j)}(D^*(x_0 A_0, A)_k)}{\Delta_C(D^*(A))} \\ &= \frac{\sum_k q_k^{(j)} \Delta_C^{(j)}(D^*(A_0, A)_k)}{\Delta_C(D^*(A))} x_0 \\ &= \frac{\sum_k w_k^{(j)} \Delta_M^{(j)}(D^*(A_0, A)_k)}{\Delta_M(D^*(A)} x_0. \end{aligned}$$

Here $q_k^{(j)}$ denotes the weight of the kth path from the vertex 0 to the vertex j and $D^*(x_0 A_0, A)_k$ is the subdigraph of the Coates digraph obtained by deleting the vertices of the kth path.

It is remarkable that the graphs introduced for treating systems of linear algebraic equations (and, especially, Mason's signal flow graph) give a better insight into the physical system under description than the corresponding system of equations does. Historically, these graphs were introduced and used intuitively, the theoretical background of them having been given later. See Section 10.1 for examples of using these techniques in electrical circuit and control theory.

6.5 Sparse Matrices

In this section we shall describe some specific features for treating systems of equations with a *sparse* matrix in which the entries are given numerically. Often problems in electrical engineering, and in engineering and science in general, lead to systems of linear equations whose matrix is sparse with entries given numerically. To a great extent, special methods of treating such matrices use graph-theoretical means [4], [11], [76], [29].

There is no strict quantitative criterion that determines when a matrix should be considered as a sparse matrix. Special procedures for treating sparse matrices include the execution of some additional operations, and therefore they are effective only if a matrix contains a sufficient number of "well-placed" entries equal to zero. Sometimes it is obvious that special techniques are not efficient. For example, if a square matrix of order 100 contains 10 zeros, it is clearly best to ignore the zeros and to solve the system by standard techniques. However, if a matrix of order 1000 contains only 4000 entries different from zero, it may be advantageous to specify the matrix by the value and the position of its nonzero entries and deal with the matrix by special methods. Roughly speaking, we should consider a matrix to be sparse if, independent of its order, each of its rows and columns contain only "a few" nonzero elements.

As indicated, a sparse matrix is stored in a computer by storing only its nonzero entries along with its row and column index. The König digraph $G(A)$ and the digraph $D(A)$ play an important role in sparse matrix techniques applied to A.

When dealing with systems of equations with a sparse matrix one would first try to split the system into subsystems, then solve each of the subsystems, and finally get the solution of the whole system from the solutions of the subsystems.

As already indicated in Section 6.1, when describing the reduction of a matrix to the reduced row-echelon form, permutation of equations and the permutation of unknowns in the equations play an important role. Permutations of rows and columns of the matrix A correspond, respectively, to permutations of equations and unknowns. Contrary to the permutations of rows and columns, which will be used in Chapter 8, it is allowed here to apply different permutations to rows and columns. In general, the matrix A is transformed by two permutation matrices P and Q, such that the new matrix has the form $A' = PAQ$. The digraph $G(PAQ)$ is obtained from the digraph $G(A)$ by independent permutations of the labels of its black vertices (action of the matrix P) and of the labels of its white vertices labels (action of the matrix Q).

Consider a system $Ax = b$, where A is a square matrix. We assume that A is nonsingular so that $Ax = b$ has a unique solution. Since A is nonsingular, $\det A \neq 0$ and hence $G(A)$ has at least one 1-factor F. We may permute the columns of A so that each edge of the 1-factor joins a black and white vertex with the same label. As a result, we get the system $A'x = b$, where each of the entries on the diagonal of the matrix A' is nonzero. As described in Section 8.1, the digraph $D(A')$ has a number m (possibly $m = 1$) of strong components. By properly relabeling rows and columns of A' (applying the same permutation to the rows and columns of A'), the matrix A' takes the following block-triangular form:

$$A' = \begin{bmatrix} A_{11} & O & \cdots & O \\ A_{21} & A_{22} & \cdots & O \\ \vdots & \vdots & \ddots & \vdots \\ A_{m1} & A_{m2} & \cdots & A_{mm} \end{bmatrix}, \qquad (6.23)$$

where $A_{11}, A_{22}, \ldots, A_{mm}$, are square blocks and all the entries above these blocks equal 0. The blocks A_{kk} correspond to the strong components of A'. (This is the Frobenius normal form of A'—actually in transposed form—as described in Section 8.1.)

Let the vectors x and b from the system $A'x = b$ be represented in the form $x^T = [x_1\, x_2 \ldots x_m]$, $b^T = [b_1\, b_2 \ldots b_m]$, where $x_1, x_2, \ldots, x_m$ and $b_1, b_2, \ldots, b_m$ are vectors of dimensions that correspond to the block sizes in (6.23). Then, solving the system $A'x = b$ is equivalent to solving the following subsystems:

$$A_{kk}x_k = b_k - \sum_{j=1}^{k-1} A_{kj}x_j, \quad k = 1, 2, \ldots, m, \qquad (6.24)$$

where we solve first the system (6.24) for $k = 1$, that is, $A_{11}x_1 = b_1$, and then solve, in order, the systems for $k = 2, \ldots, m$. Of course, if $m = 1$, we only have the original system $Ax = b$.

The digraphs $D(A)$ and $D(A')$ are not in general isomorphic. The digraph $D(A')$ depends on which 1-factor F of $G(A)$ was chosen. But what is true is that the number m of blocks in (6.23) does not depend on which 1-factor F was chosen, and the blocks A_{kk} are uniquely determined up to row and column permutations. This follows from the following observations: Similar to the proof of Theorem 4.2.10, no linear subdigraph of $D(A')$ contains edges corresponding to entries of off-diagonal blocks of A'. Therefore, for any splitting of the system into subsystems (with nonzero diagonal entries) *each* 1-factor of the König digraph $G(A')$ is a (disjoint) union of 1-factors of the König digraphs of the diagonal blocks, and therefore it does not matter which 1-factor we chose at the beginning.

Keeping in mind the above considerations, it is useful to have an algorithm for finding strong components of a digraph. Such an algorithm is given in Section 3.7 of [7]. For graph algorithms in general, one may consult see [31], [54], [39].

We are now left with the problem of solving a system that cannot be further split into smaller subsystems as described above. Thus consider a system $Ax = b$, where A is a sparse matrix, and the system cannot be split into the subsystems. We again apply the procedure for finding a reduced row-echelon form from Section 6.1 (Gaussian elimination), but its usage now has a number of special features.

In order to avoid numerical difficulties when dividing by a number close to zero, it is usually required in working with general matrices that the pivot element in each step is of maximal modulus. In sparse matrices it is required that the pivot is greater in modulus than a minimal value that is recommended in the literature for several types of problems. We shall assume that this condition is always fulfilled when making a choice of the pivoting element. This additional freedom in choosing the pivoting element enables the reduction in the number of nonzero entries by which the matrix is filled when applying EROs of type III.

Example 6.5.1 If we apply type III EROs on the system with the matrix

$$\begin{bmatrix} 1 & 2 & 3 & 4 & 5 \\ 1 & -3 & & & \\ 1 & & 2 & & \\ 1 & & & -2 & \\ 1 & & & & 2 \end{bmatrix},$$

the lower right 4×4 submatrix will be completely filled (with nonzero numbers) already in the first step of the process. However, by permuting rows and columns we get the matrix

$$\begin{bmatrix} 2 & & & & 1 \\ & -2 & & & 1 \\ & & 2 & & 1 \\ & & & -2 & 1 \\ 5 & 4 & 3 & 2 & 1 \end{bmatrix}.$$

Now the process does not lead to the appearance of new nonzero entries.

Keeping in mind the way of storing a sparse matrix in the memory of a computer and the fact that arithmetical operations are performed, only with nonzero entries, it is clear that the appearance of new nonzero elements leads to larger occupation of the memory and a longer running time of the program. It can also cause an interruption of the program execution if the space in memory is exhausted. Therefore the minimization of the number of new nonzero entries—the so-called *fill*—which appear in the process, is one of the central questions in the work with sparse matrices.

Consider the König digraph $G(A)$ of the sparse matrix A. The weight of the arc between the black vertex i and the white vertex j is a_{ij}. Let d_i^+ be the outdegree of i and d_j^- the indegree of j. If a_{ij} is chosen for the pivot element, we must produce a zero at $d_j^- - 1$ places in the j-th column, i.e., the ith row is added $d_j^- - 1$ times to other rows (previously multiplied by an appropriate number). With each such addition we add $d_i^+ - 1$ entries that are different from zero. In total, we add $(d_i^+ - 1)(d_j^- - 1)$ nonzero entries at this step. Since the matrix is sparse, as many as this number of new nonzero entries will be added to zero entries, and thus as many as $(d_i^+ - 1)(d_j^- - 1)$ new nonzero entries may appear in this step. Hence, the pivot entry is often determined by the edge (i, j) of the digraph $G(A)$ for which $(d_i^+ - 1)(d_j^- - 1)$ is minimal. Of course, in any step of the process we consider a vertex deleted subgraph of the original digraph $G(A)$.

6.6 Exercises

1. Use EROs to find all solutions of the following system of linear equations:

$$\begin{aligned} x_1 - 2x_2 + x_3 &= 3 \\ 3x_1 - 3x_2 - 2x_3 &= 6 \\ 5x_1 - 3x_2 &= 8 \end{aligned}$$

2. Use EROs to find all solutions of the system of equations $Ax = b$, where

$$A = \begin{bmatrix} 0 & 1 & 1 & 3 \\ 1 & 3 & -2 & 0 \\ 2 & 6 & -2 & 4 \end{bmatrix} \text{ and } b = \begin{bmatrix} 10 \\ 3 \\ 18 \end{bmatrix}.$$

3. Use EROs to find the inverse of the matrix

$$\begin{bmatrix} 2 & 3 & 1 \\ 1 & 0 & 2 \\ 1 & -1 & 2 \end{bmatrix}.$$

4. Find bases of the row space, column space, and null space of the matrix

$$\begin{bmatrix} -1 & -2 & 1 & 1 & 0 \\ 1 & 2 & -2 & -4 & 3 \\ 1 & 2 & 0 & 2 & 1 \\ 1 & 2 & -3 & -7 & 2 \end{bmatrix}.$$

5. Use Cramer's rule to find the solution of the following system of linear equations:

$$\begin{aligned} x_1 + 2x_2 + 3x_3 &= 1 \\ 3x_1 + 5x_2 + 8x_3 &= 1 \\ 4x_1 + 5x_2 + 10x_3 &= 1 \end{aligned}$$

6. Use Cramer's rule to find the solution of $Ax = b$, where

$$A = \begin{bmatrix} 1 & 2 & 3 & 4 \\ 4 & 1 & 2 & 3 \\ 3 & 4 & 1 & 2 \\ 2 & 3 & 4 & 1 \end{bmatrix} \text{ and } b = \begin{bmatrix} 1 \\ 0 \\ 1 \\ 0 \end{bmatrix}.$$

7. Solve the system of equations given in Example 6.3.2.

8. Draw the Coates digraph corresponding to the linear system

$$\begin{aligned} ax_2 + bx_3 \qquad\qquad &= A \\ cx_1 + dx_2 \qquad\qquad + ex_4 &= B \\ fx_1 \qquad\qquad + gx_3 + hx_4 &= 0 \\ ux_2 + vx_3 \qquad\qquad &= 0. \end{aligned}$$

Under the assumption that $acvh + bfeu - bcuh - afve \neq 0$, find its solution.

9. Using (6.24), solve the system of equations $Ax = b$, where

$$A = \begin{bmatrix} 2 & 1 & 0 & 0 \\ 1 & 1 & 0 & 0 \\ 3 & 1 & 1 & 4 \\ 1 & 0 & 1 & 2 \end{bmatrix}.$$

10. Solve the system of equations given in Example 10.1.1 in Section 10.1.

Chapter 7

Spectrum of a Matrix

In this chapter we introduce the fundamental concepts of eigenvalues and eigenvectors of a square matrix in the classical way. The eigenvalues of a matrix A of order n are roots of a polynomial, called the characteristic polynomial of A. The coefficients of this polynomial are sums of certain determinants of submatrices of A and thus can be described using digraphs as shown in Section 7.1. In Section 7.2, we give a combinatorial argument for the Cayley–Hamilton theorem, which asserts that a matrix satisfies its characteristic polynomial. The study of eigenvalues leads to the notion of similarity of matrices and this, in turn, leads to the Jordan Canonical Form of a matrix in Section 7.3. We give a highly combinatorial argument for the existence of the Jordan Canonical Form of a matrix. The chapter is concluded with Section 7.4 which describes how eigenvalues of circulants, introduced in Chapter 3, can be calculated using associated digraphs.

7.1 Eigenvectors and Eigenvalues

We begin with an important definition.

Definition 7.1.1 Let

$$A = \begin{bmatrix} a_{11} & a_{12} & \cdots & a_{in} \\ a_{21} & a_{22} & \cdots & a_{2n} \\ \vdots & \vdots & \ddots & \vdots \\ a_{n1} & a_{n2} & \cdots & a_{nn} \end{bmatrix}$$

be a matrix of order n. Let λ be a real or complex number. Then λ is an *eigenvalue* of A provided there is a nonzero column vector

$$u = \begin{bmatrix} u_1 \\ u_2 \\ \vdots \\ u_n \end{bmatrix}$$

in $\Re^n$ or $\mathcal{C}^n$ such that $Au = \lambda u$. If the eigenvalue λ is a real number, then there is a real vector u. However, if λ is a complex number, which may happen even if A is real, the vector may be a complex vector. The nonzero column vector u is called an *eigenvector* of A corresponding to its eigenvalue λ. The *eigenvalue-eigenvector* matrix equation $Au = \lambda u$ can be rewritten as

$$(\lambda I_n - A)u = 0.$$

□

Because eigenvectors of a real matrix may be nonreal, we generally take our eigenvectors in $\mathcal{C}^n$. Let u be an eigenvector of A corresponding to the eigenvalue λ. Since u is a nonzero vector, the equation $(\lambda I_n - A)u = 0$ implies that λ is an eigenvalue of A if and only if $\lambda I_n - A$ is a singular matrix; equivalently, λ is an eigenvalue of A if and only if

$$\det(\lambda I_n - A) = 0. \tag{7.1}$$

In particular, 0 is an eigenvalue of A if and only if $\det A = 0$, that is, if and only if A is a singular matrix.

Example 7.1.2 Let

$$A = \begin{bmatrix} 1 & 2 \\ 3 & 2 \end{bmatrix}.$$

A simple computation shows that

$$\det(\lambda I_2 - A) = \lambda^2 - 3\lambda - 4 = (\lambda - 4)(\lambda + 1),$$

which equals zero if and only if $\lambda = 4$ or -1. Thus A has two eigenvalues, namely, 4 and -1. We then have

$$4I_2 - A = \begin{bmatrix} 3 & -2 \\ -3 & 2 \end{bmatrix}.$$

To find an eigenvector of A corresponding to its eigenvalue 4, we need to solve the homogeneous system of two linear equations

$$\begin{aligned} 3x_1 - 2x_2 &= 0 \\ -3x_1 + 2x_2 &= 0. \end{aligned}$$

One solution is $x_1 = 2$ and $x_2 = 3$. Thus $u = [2\ 3]^T$ (and any nonzero multiple of this vector) is an eigenvector of A corresponding to $\lambda = 4$. A similar computation shows that $u = [1\ -1]^T$ (and any nonzero multiple of this vector) is an eigenvector of A corresponding to $\lambda = -1$.

Now let

$$A = \begin{bmatrix} 1 & 2 \\ -1 & -1 \end{bmatrix}.$$

Then

$$\det(\lambda I_2 - A) = \lambda^2 + 1 = (\lambda - i)(\lambda + i).$$

Hence the eigenvalues of A are $\pm i$. □

It follows from the definition of the determinant given in Chapter 4 that if A is a matrix of order n, then $p_A(\lambda) = \det(\lambda I_n - A)$ is a polynomial in λ of degree n, called the *characteristic polynomial*[1] of A. Since $p_A(\lambda)$ is a polynomial of degree n, and since

---

[1]An eigenvalue is also called a *characteristic value* and an eigenvector is also called a *characteristic vector*.

a polynomial of degree n has n roots (possibly complex numbers even if A is a real matrix), counting multiplicities, the matrix A has n eigenvalues $\lambda_1, \lambda_2, \ldots, \lambda_n$. These n eigenvalues comprise the *spectrum* of A.

Example 7.1.3 Let

$$A = \begin{bmatrix} 2 & 0 & 0 & 0 \\ 0 & 5 & 0 & 0 \\ 0 & 0 & 2 & 0 \\ 0 & 0 & 0 & 5 \end{bmatrix}.$$

Then $p_A(\lambda) = \det(\lambda I_4 - A) = (\lambda-2)^2(\lambda-5)^2$. Thus the eigenvalues of A are $2, 2, 5, 5$. More generally, the eigenvalues of a diagonal matrix, indeed a triangular matrix, of order n are its n diagonal entries. □

From Corollary 4.3.2 we obtain that the characteristic polynomial of a matrix A of order n is given by

$$\det(\lambda I_n - A) = \sum_{p=0}^{n} (-1)^{n-p} c_{n-p} \lambda^p, \tag{7.2}$$

where

$$c_{n-p} = \sum_{1 \leq j_1 < j_2 < \cdots < j_{n-p} \leq n} \det A[\{j_1, j_2, \ldots, j_{n-p}\}, \{j_1, j_2, \ldots, j_{n-p}\}],$$

the sum of the determinants of all the principal submatrices of A of order $n-p$. In particular, we have that the coefficient of λ^{n-1} is

$$c_1 = a_{11} + a_{22} + \cdots + a_{nn} = \mathrm{tr} A,$$

and

$$c_n = \det A.$$

Example 7.1.4 Let $A = [a_{ij}]$ be a square matrix of order 3. Then

$$\lambda I_3 - A = \begin{bmatrix} \lambda - a_{11} & -a_{22} & -a_{13} \\ -a_{21} & \lambda - a_{22} & -a_{23} \\ -a_{31} & -a_{32} & \lambda - a_{33} \end{bmatrix}.$$

Thus the characteristic polynomial of A is

$$p_A(\lambda) = \lambda^3 - (\mathrm{tr}A)\lambda^2 + c_2\lambda - \det A,$$

where

$$c_2 = (a_{11}a_{22} - a_{12}a_{21}) + (a_{11}a_{33} - a_{13}a_{31}) + (a_{22}a_{23} - a_{23}a_{32}),$$

a sum that can be rewritten as

$$c_2 = \det A[\{1,2\},\{1,2\}] + \det A[\{1,3\},\{1,3\}] + \det A[\{2,3\},\{2,3\}].$$

Thus c_2 is the sum of the determinants of the three principal submatrices of A of order 2. □

Example 7.1.5 Let

$$A = \begin{bmatrix} 1 & 1 & 1 & 1 \\ 1 & 1 & 1 & 1 \\ 1 & 1 & 1 & 1 \\ 1 & 1 & 1 & 1 \end{bmatrix}.$$

Since the determinant of a matrix of order at least 2, each of whose entries equals 1 is 0, computing the characteristic polynomial of A using formula (7.2), we get that $c_1 = 4$, $c_2 = c_3 = c_4 = 0$, and hence

$$p_A(\lambda) = \lambda^4 - 4\lambda^3 = \lambda^3(\lambda - 4).$$

Hence the eigenvalues of A are $4, 0, 0, 0$. More generally, the eigenvalues of the matrix of all 1's of order n are $n, 0, \ldots, 0$ where there are $n-1$ 0's. □

Example 7.1.6 Let a matrix A of order n be the direct sum of square matrices, say,

$$A = A_1 \oplus A_2 \oplus A_3$$

of orders n_1, n_2, and n_3, respectively. Then

$$\lambda I_n - A = (\lambda I_{n_1} - A_1) \oplus (\lambda I_{n_2} - A_2) \oplus (\lambda I_{n_3} - A_3),$$

and hence

$$\det(\lambda I_n - A) = \det(\lambda I_{n_1} - A_1)\det(\lambda I_{n_2} - A_2)\det(\lambda I_{n_3} - A_3),$$

and the characteristic polynomial of A is the product of the characteristic polynomials of $A_1, A_2,$ and A_3. Thus the spectrum of A is obtained by putting the spectra of $A_1, A_2,$ and A_3 together. □

Let A be a matrix of order n with eigenvalues $\lambda_1, \lambda_2, \ldots, \lambda_n$. Because the eigenvalues are the n roots of the characteristic polynomial $p_A(\lambda)$, which has leading coefficient 1, we have

$$\begin{aligned} p_A(\lambda) &= (\lambda - \lambda_1)(\lambda - \lambda_2)\cdots(\lambda - \lambda_n) \\ &= \lambda^n - (\lambda_1 + \lambda_2 + \cdots + \lambda_n)\lambda^{n-1} + \cdots + (-1)^n \lambda_1\lambda_2\cdots\lambda_n. \end{aligned}$$

Comparing with the coefficients of the characteristic polynomial as given in (7.2), we see that the trace of A is the sum of its n eigenvalues:

$$c_1 = \mathrm{tr}A = \lambda_1 + \lambda_2 + \cdots + \lambda_n,$$

and the determinant of A is the product of these eigenvalues:

$$c_n = \lambda_1\lambda_2\cdots\lambda_n.$$

Because $\lambda I_n - A^T = (\lambda I_n - A)^T$, and a matrix and its transpose have the same determinant,

$$p_{A^T}(\lambda) = \det(\lambda I_n - A^T) = \det((\lambda I_n - A)^T) = \det(\lambda I_n - A) = p_A(\lambda),$$

that is, A and A^T have the same characteristic polynomial and hence the same eigenvalues.

Definition 7.1.7 Let $A = [a_{ij}]$ be a matrix of order n and let λ be an eigenvalue of A. The *algebraic multiplicity* of λ is its multiplicity as a root of the characteristic polynomial $\det(\lambda I_n - A)$. Thus the multiplicity of an eigenvalue of A is an integer between 1 and n. The *eigenspace* of A corresponding to λ is defined to be

$$V_\lambda(A) = \{x \in C^n : (\lambda I_n - A)x = 0\}.$$

The eigenspace of A corresponding to λ is the null space of the matrix $\lambda I_n - A$, and so is a subspace of $\mathcal{C}^n$. (If all n eigenvalues of A are real numbers, then we can take the eigenspace to be a subspace of $\Re^n$. But if A has a complex eigenvalue, then the eigenspaces are taken to be subspaces of $\mathcal{C}^n$.) The eigenspace consists of the zero vector and all the eigenvectors of A corresponding to λ, and thus has dimension at least 1. The *geometric multiplicity* of λ is the dimension of the eigenspace $V_\lambda(A)$ and thus equals $n - \mathrm{r}(\lambda I_n - A)$. □

Example 7.1.8 First consider the identity matrix I_n. Each of its n eigenvalues equals 1, and the eigenspace $V_1(I_n)$ is all of $\Re^n$. Thus both the algebraic and geometric multiplicities of 1 equal n. More generally, if D is a diagonal matrix of order n with diagonal entries $d_1, d_2, \ldots, d_n$, then its eigenvalues are $d_1, d_2, \ldots, d_n$ and the geometric multiplicity of an eigenvalue equals its algebraic multiplicity. For example, let

$$D = \begin{bmatrix} 4 & 0 & 0 & 0 & 0 \\ 0 & 4 & 0 & 0 & 0 \\ 0 & 0 & 7 & 0 & 0 \\ 0 & 0 & 0 & 7 & 0 \\ 0 & 0 & 0 & 0 & 7 \end{bmatrix}.$$

Then the eigenvalues of D are $4, 4, 7, 7, 7$, so that 4 is an eigenvalue with algebraic multiplicity 2 and 7 is an eigenvalue with algebraic multiplicity 3. We then have

$$4I_5 - D = \begin{bmatrix} 0 & 0 & 0 & 0 & 0 \\ 0 & 0 & 0 & 0 & 0 \\ 0 & 0 & -3 & 0 & 0 \\ 0 & 0 & 0 & -3 & 0 \\ 0 & 0 & 0 & 0 & -3 \end{bmatrix},$$

a matrix of rank 3. Hence the dimension of the null space of $4I_5 - D$, that is, the geometric multiplicity of 4, equals 2. In a similar way, we see that the geometric multiplicity of 7 is 3.

Now consider the matrix T_n that has 1's on and above the main diagonal. For instance, if $n = 4$, then

$$T_4 = \begin{bmatrix} 1 & 1 & 1 & 1 \\ 0 & 1 & 1 & 1 \\ 0 & 0 & 1 & 1 \\ 0 & 0 & 0 & 1 \end{bmatrix}.$$

The characteristic polynomial of T_n is $p_{T_n}(\lambda) = (\lambda - 1)^n$ and so 1 is an eigenvalue of T_n with algebraic multiplicity equal to n. The eigenspace $V_1(T_n)$ consists of all vectors $u = [u_1\ u_2\ \ldots u_n]^T$ such that $(I_n - T_n)u = 0$. The matrix $I_n - T_n$ clearly has rank equal to $n - 1$ and hence the dimension of its null space equals 1. Thus the geometric multiplicity of 1 equals 1. We conclude from this example that the algebraic and geometric multiplicities of an eigenvalue need not be equal and indeed may be quite different. These facts play an important role later in this chapter. □

We conclude this section by showing that each of the eigenvalues of a real symmetric are real numbers.

Theorem 7.1.9 *Let $A = [a_{ij}]$ be a real symmetric of order n. Then each of the n eigenvalues of A is a real number.*

Proof. We use the dot product of $\mathcal{C}^n$ as reviewed in Section 1.5. Let λ be any eigenvalue of A, and let x be an eigenvector (a nonzero vector in $\mathcal{C}^n$) for λ:

$$Ax = \lambda x. \tag{7.3}$$

For a matrix B, let B^H denote the matrix obtained from B by replacing each of its entries by its complex conjugate and transposing (or, equivalently, transposing and then conjugating each entry). We have $(BC)^H = C^H B^H$, since $(BC)^T = C^T B^T$, and $\overline{a+b} = \overline{a} + \overline{b}$ and $\overline{ab} = \overline{a}\overline{b}$. Since A is real and symmetric, $A^H = A$. Multiplying (7.3) by x^H we get $x^H Ax = \lambda x^H x$. We also have from (7.3) that

$$x^H Ax = (x^H A)x = (Ax)^H x = (\lambda x)^H x = \overline{\lambda} x^H x.$$

Thus $\lambda x^H x = \overline{\lambda} x^H x$. Because x is not a zero vector, $x^H x \neq 0$, and hence $\lambda = \overline{\lambda}$ and λ is a real number. □

7.2 The Cayley–Hamilton Theorem

The Cayley–Hamilton theorem asserts the rather surprising result that a square matrix satisfies its characteristic polynomial. This is a very important theorem with theoretical consequences and with applications in physics and engineering disciplines. Before giving a precise statement of the Cayley–Hamilton theorem, we introduce some concepts that we will use in its verification.

Definition 7.2.1 Let D be a digraph with vertices $1, 2, \ldots, n$, and let i and j be vertices of D. We recall that a 1-connection $D[i \to j]$ of vertex i to vertex j is a spanning subdigraph of D consisting of a path from i to j and a possibly empty collection of pairwise vertex disjoint cycles having no vertex in common with the path. The total number of edges in a 1-connection equals $n-1$. A *quasi-1-connection* $D[i \to j]^*$ from i to j is defined like a 1-connection except that the path from i to j is replaced with a walk from i to j of length at most n where the walk may intersect the cycles and where the total number of edges in the walk and cycles is to equal n.[2] Thus a quasi-1-connection $D[i \to j]^*$ is a pair consisting of a walk γ from i to j and a possibly empty collection C of pairwise vertex-disjoint cycles, where the total number of edges in the walk and cycles equals n. The *weight* $w(D[i \to j]^*)$ of a quasi-1-connection $D[i \to j]^*$ is the product of the weights of all its edges. The number of cycles in the quasi-1-connection is the number of cycles in C, and this number is denoted by $c(D[i \to j]^*)$. □

Theorem 7.2.2 *(Cayley–Hamilton theorem) Let $A = [a_{ij}]$ be a matrix of order n, and let*

$$p_A(\lambda) = \lambda^n - c_1\lambda^{n-1} + c_2\lambda^{n-2} - \cdots + (-1)^k c_{n-k}\lambda^k + \cdots + (-1)^n c_n$$

be the characteristic polynomial of A. Then $p(A) = O$, that is,

$$A^n - c_1A^{n-1} + c_2A^{n-2} - \cdots + (-1)^k c_{n-k}A^k + \cdots + (-1)^n c_n I_n = O. \tag{7.4}$$

[2] Unlike a 1-connection, the total number of edges in a quasi-1-connection is not determined by its walk and cycles.

Before proving this theorem we give an example.

Example 7.2.3 Let

$$A = \begin{bmatrix} 1 & 2 & 0 \\ -1 & 3 & 1 \\ 2 & 1 & 1 \end{bmatrix}.$$

A simple calculation shows that

$$p_A(\lambda) = \lambda^3 - 5\lambda^2 + 8\lambda - 8.$$

We calculate that

$$A^2 = \begin{bmatrix} -1 & 8 & 2 \\ -2 & 8 & 4 \\ 3 & 8 & 2 \end{bmatrix} \text{ and } A^3 = \begin{bmatrix} -5 & 24 & 10 \\ -2 & 24 & 12 \\ -1 & 32 & 10 \end{bmatrix}.$$

Substituting into the characteristic polynomial of A we obtain

$$\begin{bmatrix} -5 & 24 & 10 \\ -2 & 24 & 12 \\ -1 & 32 & 10 \end{bmatrix} - 5\begin{bmatrix} -1 & 8 & 2 \\ -2 & 8 & 4 \\ 3 & 8 & 2 \end{bmatrix} +$$

$$8\begin{bmatrix} 1 & 2 & 0 \\ -1 & 3 & 1 \\ 2 & 1 & 1 \end{bmatrix} - 8\begin{bmatrix} 1 & 0 & 0 \\ 0 & 1 & 0 \\ 0 & 0 & 1 \end{bmatrix} = O.$$

□

Proof of Theorem 7.2.2. We have to show that each entry of the matrix $p_A(A)$ as given in (7.4) equals 0. The coefficient c_{n-k} of λ^k in the characteristic polynomial equals the sum of all the determinants of the principal submatrices of A of order $n-k$. It follows from the definition of the determinant given in Chapter 4 that c_{n-k} equals

$$(-1)^{n-k} \sum_L (-1)^{c(L)} w(L),$$

where the summation extends over all linear subdigraphs of the Coates digraph $D^*(A)$ having $n-k$ vertices and where $c(L)$ is the number of cycles of L. From Chapter 3 we know that the entry in position (i,j) of A^k equals the sum of the weights of all walks of length k from vertex i to vertex j.

Therefore the entry in position (i,j) of $(-1)^{n-k}c_{n-k}A^k$ equals

$$\sum_{D[i\to j]_k^*} (-1)^{c(L)} w(D[i \to j]_k^*),$$

where the summation extends over all quasi-1-connections $D[i \to j]_k^*$ whose walk γ from i to j has length k. We thus conclude that the entry in position (i,j) of $p_A(A)$ equals

$$\sum_{D[i\to j]^*} (-1)^{c(D[i\to j]^*)} w(D[i \to j]^*), \tag{7.5}$$

where the summation now extends over all quasi-connections from i to j. (Recall that the walk in a quasi-1-connection has length at most n.)

We now show how to pair up the terms[3] in (7.5) so that the sum of the terms of each pair equals zero. The total number of edges in a quasi-1-connection consisting of a walk γ and a collection C of pairwise vertex-disjoint cycles equals n. Thus either the walk γ is not a path (and so contains a cycle) or the walk γ meets a vertex of one of the cycles in C. We proceed along the walk γ until we first (a) revisit a vertex, or (b) visit a vertex of one of the cycles π in C, whichever comes first (note that these two events cannot occur simultaneously). In case (a), the walk contains a cycle and we remove this cycle from γ and add it to C to create a new quasi-1-connection with one more cycle but the same weight. In case (b) we remove the cycle π from C and add it to the walk γ to make it larger. In each case the number of cycles changes by 1, so the sign in (7.5) changes but the weight of the quasi-1-connection does not change. This process is reversible, leading us to conclude that the sum in (7.5) equals zero. Hence $p_A(A) = O$, as was to be proved. □

We have given a proof of the classical Cayley–Hamilton theorem that illustrates that it is really a theorem about weighted digraphs; this was first noted by Rutherford [70] (see also [75], [81] and [7]).

[3]That is, establish an *involution*.

7.3 Similar Matrices and the JCF

An m by n matrix $A = [a_{ij}]$ represents the linear transformation T from an n-tuple space F^n to an m-tuple space F^m defined by multiplication by A as follows: If $x = [x_1\ x_2\ \cdots\ x_n]^T$ is an n-tuple, then

$$T(x) = A \begin{bmatrix} x_1 \\ x_2 \\ \vdots \\ x_n \end{bmatrix} = \begin{bmatrix} y_1 \\ y_2 \\ \vdots \\ y_m \end{bmatrix}.$$

Here the vectors x and y are given in terms of their coordinates with respect to the standard (ordered) basis $\eta = (e_1, e_2, \ldots, e_n)$ of F^n and the standard (ordered) basis $\eta' = (e'_1, e'_2, \ldots, e'_m)$ of F^m, respectively. If we choose a different (ordered) basis $\alpha = (u^{(1)}, u^{(2)}, \ldots, u^{(n)})$ for F^n and a different (ordered) basis $\beta = (w^{(1)}, w^{(2)}, \ldots, w^{(m)})$ for F^m, then, with respect to coordinates relative to these bases, a different matrix will represent the linear transformation T. A vector u in F^n can be uniquely represented as a linear combination of the vectors in α:

$$u = p_1 u^{(1)} + p_2 u^{(2)} + \cdots + p_n u^{(n)}$$

and has *coordinate vector*

$$[u]_\alpha = \begin{bmatrix} p_1 \\ p_2 \\ \vdots \\ p_n \end{bmatrix}$$

with respect to the basis α. Similarly, the vector $w = T(u)$ in F^m can be uniquely represented as a linear combination of the vectors in β:

$$T(u) = q_i w^{(1)} + q_2 w^{(2)} + \cdots + q_m w^{(m)}$$

and has *coordinate vector*

$$[T(u)]_\beta = \begin{bmatrix} q_1 \\ q_2 \\ \vdots \\ q_m \end{bmatrix}$$

with respect to the basis β. Let $R = [r_{ij}]$ be the n by n matrix whose column vectors are the coordinate vectors

$$[u^{(1)}]_\eta, [u^{(2)}]_\eta, \ldots, [u^{(n)}]_\eta,$$

and let $S = [s_{ij}]$ be the m by m matrix whose column vectors are the coordinate vectors

$$[e'_1]_\beta, [e'_2]_\beta, \ldots, [e'_m]_\beta.$$

A straightforward calculation shows that

$$\begin{aligned}
T(u) &= T\left(\sum_{i=1}^{n} p_i u^{(i)}\right) \\
&= \sum_{i=1}^{n} p_i T\left(u^{(i)}\right) \\
&= \sum_{i=1}^{n} p_i T\left(\sum_{j=1}^{n} r_{ji} e_j\right) \\
&= \sum_{i=1}^{n} p_i \sum_{j=1}^{n} r_{ji} T(e_j) \\
&= \sum_{i=1}^{n} p_i \sum_{j=1}^{n} r_{ji} \left(\sum_{k=1}^{m} a_{kj} e'_k\right) \\
&= \sum_{i=1}^{n} p_i \sum_{j=1}^{n} r_{ji} \sum_{k=1}^{m} a_{kj} \sum_{l=1}^{m} s_{lk} w^{(l)} \\
&= \sum_{l=1}^{m} \left(\sum_{i=1}^{n} \sum_{k=1}^{m} \sum_{j=1}^{n} s_{lk} a_{kj} r_{ji} p_i\right) w^{(l)}.
\end{aligned}$$

Thus

$$[T(u)]_\beta = SAR[u]_\alpha.$$

In the special case that $n = m$ and $\alpha = \beta$, the matrix S equals R^{-1}, and hence we get

$$[T(u)]_\alpha = R^{-1}AR[u]_\alpha.$$

Hence, *the matrix of the linear transformation* $T : \Re^n \to \Re^n$ *with respect to a basis* α *of* $\Re^n$ *equals* $R^{-1}AR$*, where* A *is the matrix of*

T with respect to the standard basis η of $\Re^n$. This motivates the following definition.

Definition 7.3.1 Let A and B be square matrices of the same order n. Then B is *similar* to A provided there is an invertible matrix X such that $B = X^{-1}AX$. If B is similar to A, then we write $B \sim A$. □

The relation of similarity on matrices of order n satisfies three important properties:

(R) *reflexive property*: $B \sim B$ for all B: This is because we may take $X = I_n$ in the definition of similarity.

(S) *symmetric property*: If $B \sim A$, then also $A \sim B$: If $B \sim A$, then $B = X^{-1}AX$ for some invertible matrix X, and then $A = XBX^{-1} = (X^{-1})^{-1}A(X^{-1}) = Y^{-1}BY$, where $Y = X^{-1}$.

The symmetry property implies that we may simply say that matrices A and B are similar rather than say that A is similar to B.

(T) *transitive property*: If $A \sim B$ and $B \sim C$, then $A \sim C$.

If $A \sim B$ and $B \sim C$, then $A = X^{-1}BX$ and $B = Y^{-1}CY$ for some invertible matrices X and Y. Hence

$$A = X^{-1}Y^{-1}CYX = (YX)^{-1}C(YX) = Z^{-1}CZ,$$

where $Z = YX$.

The properties of reflexive, symmetric, and transitive are the three properties defining an *equivalence relation*. An equivalence relation on a set always partitions the set into equivalence classes whereby two elements in the same class are equivalent and two elements in different classes are not equivalent. Thus similarity defines an equivalence relation on the set of square matrices of order n and partitions the matrices into *similarity classes*, that is, classes such that matrices in the same class are similar and those in different classes are not.

In the next theorem we collect several elementary properties of similar matrices.

Theorem 7.3.2 *Let A and B be similar matrices of order n. Then the following properties hold:*

(i) $\det A = \det B$.

(ii) *The rank of A equals the rank of B.*

(iii) *A and B have the same characteristic polynomial and the same spectrum. Thus the algebraic multiplicity of eigenvalues is the same for A and for B.*

(iv) *Let B be similar to A with $B = X^{-1}AX$, and let λ be an eigenvalue of A with corresponding eigenvector u. Then $X^{-1}u$ is an eigenvector of B corresponding to its eigenvalue λ. Likewise, if v is an eigenvector of B corresponding to eigenvalue λ, then Xv is an eigenvector of A corresponding to eigenvalue λ. Thus*

$$V_\lambda(B) = \{X^{-1}u : u \in V_\lambda(A)\},$$

and the geometric multiplicities of the eigenvalues are the same for A and for B.

Proof. Assume that $B = X^{-1}AX$. Then

$$\begin{aligned} \det B &= \det(X^{-1}AX) = \det X^{-1} \det A \det X \\ &= (\det X)^{-1} \det A \det X = \det A. \end{aligned}$$

Thus (i) holds. Assertion (ii) follows from the fact that multiplying a matrix by an invertible matrix, thus by a product of elementary matrices, does not change its rank. For (iii) we calculate that

$$\begin{aligned} \det(\lambda I_n - B) &= \det(\lambda I_n - X^{-1}AX) = \det(X^{-1}(\lambda I_n - A)X) \\ &= \det X^{-1} \det(\lambda I_n - A) \det X = \det(\lambda I_n - A). \end{aligned}$$

For property (iv) we simply calculate that

$$B(X^{-1}u) = (X^{-1}AX)(X^{-1}u) = X^{-1}Au = X^{-1}\lambda u = \lambda(X^{-1}u)$$

and note that since u is a nonzero vector, so is $X^{-1}u$. Let λ be an eigenvalue of A, and thus for B, and let $Bv = \lambda v$. A similar calculation can be made with A and Xv. Thus $V_\lambda(B) = \{X^{-1}u : u \in V_\lambda(A)\}$. Because X^{-1} is a nonsingular matrix, it follows that $V_\lambda(A)$ and $V_\lambda(B)$ have the same dimension. □

Example 7.3.3 The conditions (i)–(iii) do not guarantee that the matrices A and B are similar. For example, let

$$A = I_2 = \begin{bmatrix} 1 & 0 \\ 0 & 1 \end{bmatrix} \text{ and } B = \begin{bmatrix} 1 & 1 \\ 0 & 1 \end{bmatrix}.$$

Then A and B have determinant equal to 1, have rank equal to 2, and have characteristic polynomial equal to $(\lambda - 1)^2$. But A and B are not similar since the only matrix similar to I_2 is I_2 (in general, $X^{-1}I_nX = I_n$). The geometric multiplicity of 1 is 2 for A and 1 for B.

If we also assume that the geometric multiplicities of the corresponding eigenvalues of A and B are equal, we still cannot conclude that A and B are similar. For example, let

$$A = \begin{bmatrix} 0 & 1 & 0 & 0 \\ 0 & 0 & 0 & 0 \\ 0 & 0 & 0 & 1 \\ 0 & 0 & 0 & 0 \end{bmatrix} \text{ and } B = \begin{bmatrix} 0 & 0 & 0 & 0 \\ 0 & 0 & 1 & 0 \\ 0 & 0 & 0 & 1 \\ 0 & 0 & 0 & 0 \end{bmatrix}.$$

Then 0 is an eigenvalue of A and of B with algebraic multiplicity 4 and geometric multiplicity 2. By calculation we see that $A^2 = O$ but $B^2 \neq O$. But if $B = X^{-1}AX$, then $B^2 = X^{-1}A^2X = X^{-1}OX = O$. So B is not similar to A. □

Example 7.3.4 Let A and B be two square matrices of the same order n. Suppose that the digraphs $D(A)$ and $D(B)$ are isomorphic. Then there is a permutation matrix P such that $B = P^TAP$.

Since the inverse of a permutation matrix equals its transpose, we can rewrite this equation as $B = P^{-1}AP$, and hence A and B are similar via a permutation matrix. The converse also holds. Thus the isomorphism of digraphs is equivalent to similarity via a permutation matrix. A similar conclusion holds if we consider the Coates digraphs $D^*(A)$ and $D^*(B)$. On the other hand, if the König digraphs $G(A)$ and $G(B)$ are isomorphic, the matrices A and B are connected by the relation $A = QBP$, where P and Q are permutation matrices. Here the permutation matrix Q permutes the black vertices of the digraph $G(B)$ and the permutation matrix P independently permutes the white vertices. If the matrices A and B are similar, requiring $Q = P^{-1}$, then the digraphs $G(A)$ and $G(B)$ are isomorphic where the isomorphism φ preserves the initial pairing of the black and white vertices. This means that if φ takes the black vertex i of $G(B)$ to the black vertex j of $G(A)$, then φ takes the white vertex i of $G(B)$ to the white vertex j of $G(A)$. Hence, the digraph $G(A)$ can be obtained from the digraph $G(B)$ by a permutation of the pairs of black and white vertices. □

Definition 7.3.5 Let A be a square matrix of order n. Then an *elementary similarity* of A is a matrix $B = E^{-1}AE$, where E is an elementary matrix. Since there are three types of elementary matrices, there are three types of elementary similarities:

(i) (*elementary permutation similarity*)

$$I_n(i,j)AI_n(i,j)^{-1} = I_n(i,j)AI_n(i,j),$$

in which rows i and j are switched and columns i and j are switched $(i \neq j)$. In particular, the (i,i) and (j,j) entries of the main diagonal of A are switched and the (i,j) and (j,i) entries are switched in this type of elementary similarity.

(ii) (*elementary diagonal similarity*)

$$I_n(c\cdot i)AI_n(c\cdot i)^{-1} = I_n(c\cdot i)AI_n(1/c\cdot i),$$

in which row i is multiplied by c and column i is multiplied by $1/c$ $(c \neq 0)$. There is no change in the entries of the main diagonal of A in this type of similarity.

(iii) (*elementary combination similarity*)

$$I_n(c \cdot i + j) A I_n(c \cdot i + j)^{-1} = I_n(c \cdot i + j) A I_n(-c \cdot i + j),$$

in which c times row i is added to row j and $-c$ times column j is added to column i $(i \neq j)$.

Since a matrix is invertible if and only if it is a product of elementary matrices, we conclude that *A is similar to a matrix B if and only if B can be obtained from A by a sequence of elementary similarities.* □

We now consider the question of how simple a matrix we can find in each similarity class. Here by simple we mean a matrix B for which the structure of the digraph $D(B)$ associated with the nonzero off-diagonal entries of B is simple with very few edges (so we ignore all loops of the digraph $D(A)$). Let us denote by $\widehat{D}(B)$ the digraph obtained from $D(B)$ by removing all loops. Thus, if B is a diagonal matrix, $\widehat{D}(B)$ is a digraph with no edges and thus consists of a collection of isolated vertices. This is the simplest possible structure but one that cannot always be attained. For example, for the matrix

$$B = \begin{bmatrix} 1 & 1 \\ 0 & 1 \end{bmatrix}$$

from Example 7.3.3, the digraph $\widehat{D}(B)$ consists of two vertices and an edge from one to the other. The matrix B cannot be similar to a diagonal matrix, as that diagonal matrix would have to be I_2 and this has already been ruled out in Example 7.3.3. Since similar matrices have the same spectrum, if a matrix A is similar to a diagonal matrix B, then the entries on the main diagonal of B are the n eigenvalues of A.

A matrix is *diagonalizable* provided it is similar to a diagonal matrix. In the next theorem we give a characterization, in terms of eigenvectors, of diagonalizable matrices.

Theorem 7.3.6 *Let A be a square matrix of order n. Then A is diagonalizable if and only if A has n linearly independent eigenvectors, that is, there is a basis of C^n (or $\Re^n$ if A only has real eigenvalues) consisting of eigenvectors of A.*

Proof. First suppose that $u^{(1)}, u^{(2)}, \ldots, u^{(n)}$ are n linearly independent eigenvectors of A with

$$Au^{(i)} = \lambda_i u^{(i)} \quad (i = 1, 2, \ldots, n). \tag{7.6}$$

Let U be the matrix whose columns are $u^{(1)}, u^{(2)}, \ldots, u^{(n)}$, respectively. Then U is an invertible matrix and the equations in (7.6) can be written as the one matrix equation $AU = U\Lambda$, where Λ is the diagonal matrix

$$\begin{bmatrix} \lambda_1 & 0 & \cdots & 0 \\ 0 & \lambda_2 & \cdots & 0 \\ \vdots & \vdots & \ddots & \vdots \\ 0 & 0 & \cdots & \lambda_n \end{bmatrix}.$$

Since U is invertible, we have $U^{-1}AU = \Lambda$, and A is similar to a diagonal matrix.

Conversely, if A is similar to the diagonal matrix Λ, then there is an invertible matrix P such that $P^{-1}AP = \Lambda$ and so $AP = P\Lambda$. Since P is invertible, we see that the columns of P are n linearly independent eigenvectors of A. □

We shall find a simple matrix in each similarity class in steps. We first prove a theorem that can be rephrased to say that a matrix is similar to a matrix T whose digraph $\widehat{D}(T)$ is acyclic, that is, has no cycles; indeed, the vertices can be ordered from top to bottom with all edges pointing downward.

Theorem 7.3.7 *The matrix A of order n is similar to an upper triangular matrix T. The diagonal entries of T are the n eigenvalues of A, and T can be chosen so that these eigenvalues appear on its main diagonal in any specified order $\lambda_1, \lambda_2, \ldots, \lambda_n$.*

Proof. The proof is by induction on n. If $n = 1$, there is nothing to prove as a square matrix of order 1 is upper triangular. Let λ_1 be any eigenvalue of A with corresponding eigenvector u. Since u is not the zero vector, u can be extended to a basis $u^{(1)} = u, u^{(2)}, \ldots, u^{(n)}$. Let U be the matrix whose columns are

$u^{(1)}, u^{(2)}, \ldots, u^{(n)}$, respectively. Then U is an invertible matrix. The equations $U^{-1}U = I_n$ and $Au^{(1)} = \lambda_1 u^{(1)}$ imply that

$$U^{-1}u^{(1)} = \begin{bmatrix} 1 \\ 0 \\ \vdots \\ 0 \end{bmatrix}$$

and

$$U^{-1}AU = \begin{bmatrix} \lambda_1 & \alpha_1 \\ O & A_1 \end{bmatrix}, \tag{7.7}$$

where α_1 is a 1 by $n-1$ matrix and A_1 is a square matrix of order $n-1$. The matrix in (7.7) is similar to A and hence the eigenvalues of A_1 are $\lambda_2, \ldots, \lambda_n$. By induction, there is an invertible matrix W_1 of order $n-1$ such that $W_1^{-1}A_1W_1 = T_1$, where T_1 is an upper triangular matrix with $\lambda_2, \ldots, \lambda_n$ on its main diagonal in this order. Define a partitioned matrix of order n by

$$W = \begin{bmatrix} 1 & O \\ O & W_1 \end{bmatrix}.$$

The matrix W is invertible with

$$W^{-1} = \begin{bmatrix} 1 & O \\ O & W_1^{-1} \end{bmatrix}.$$

Then UW is an invertible matrix, and using block multiplication, we get that

$$(UW)^{-1}A(UW) = W^{-1}(U^{-1}AU)W = \begin{bmatrix} \lambda_1 & \beta \\ O & T_1 \end{bmatrix},$$

an upper triangular matrix with diagonal entries $\lambda_1, \lambda_2, \ldots, \lambda_n$. Hence the theorem holds by induction. □

Corollary 7.3.8 *Let A be a matrix of order n with eigenvalues $\lambda_1, \lambda_2, \ldots, \lambda_n$. Let k be a positive integer. Then the eigenvalues of*

A^k are $\lambda_1^k, \lambda_2^k, \ldots, \lambda_n^k$. More generally, if $q(x) = c_k x^k + c_{k-1}x^{k-1} + \cdots + c_1 x + c_0$ is a polynomial, then the eigenvalues of

$$q(A) = c_k A^k + c_{k-1} A^{k-1} + \cdots + c_1 A + c_0 I_n$$

are $q(\lambda_1), q(\lambda_2), \ldots, q(\lambda_n)$. In addition, if A is invertible, then the eigenvalues of A^{-1} are $\lambda_1^{-1}, \lambda_2^{-1}, \ldots, \lambda_n^{-1}$.

Proof. By Theorem 7.3.7 there is an invertible matrix Q such that $A = Q^{-1}TQ$, where T is an upper triangular matrix with $\lambda_1, \lambda_2, \ldots, \lambda_n$ on its main diagonal. Then

$$A^k = (Q^{-1}TQ)^k = Q^{-1}T^kQ,$$

and hence A^k is similar to T^k. Since T^k is an upper triangular matrix whose entries on the main diagonal are $\lambda_1^k, \lambda_2^k, \ldots, \lambda_n^k$, the eigenvalues of A^k are $\lambda_1^k, \lambda_2^k, \ldots, \lambda_n^k$. More generally,

$$q(A) = q(Q^{-1}TQ) = Q^{-1}q(T)Q,$$

and it follows that the eigenvalues of $q(A)$ are $q(\lambda_1), q(\lambda_2), \ldots, q(\lambda_n)$.

If A is invertible, then

$$A^{-1} = (Q^{-1}TQ)^{-1} = QT^{-1}Q^{-1},$$

where T^{-1} is an upper triangular matrix similar to A^{-1}, whose entries on the main diagonal are $\lambda_1^{-1}, \lambda_2^{-1}, \ldots, \lambda_n^{-1}$. Hence these are the n eigenvalues of A^{-1}. □

Our next goal is to show that the matrix T in Theorem 7.3.7, for which $\widehat{D}(T)$ is acyclic, is similar to a matrix J such that the digraph $\widehat{D}(J)$ is a collection of vertex-disjoint paths (and so certainly acyclic).

Definition 7.3.9 Let k be a positive integer. A matrix of order k of the form

$$J_k(\mu) = \begin{bmatrix} \mu & 1 & \cdots & 0 & 0 \\ 0 & \mu & \cdots & 0 & 0 \\ \vdots & \vdots & \ddots & \vdots & \vdots \\ 0 & 0 & \cdots & \mu & 1 \\ 0 & 0 & \cdots & 0 & \mu \end{bmatrix}$$

with μ's on the main diagonal and 1's on the superdiagonal (the diagonal immediately above the main diagonal) is called a *Jordan block* of order k. If $k = 1$, then $J_1(\mu)$ is a matrix of order 1 whose unique entry equals μ. The Jordan block $J_k(\mu)$ has a characteristic polynomial $(\lambda - \mu)^k$ and hence has μ as an eigenvalue of algebraic multiplicity k. The geometric multiplicity of μ equals the dimension of the eigenspace $V_\mu(J_k)$, and this equals $k - \mathrm{r}(\mu I_k - J_k) = k - (k-1) = 1$. Notice that the digraph $\widehat{D}(J_k)$ is a path with k vertices.

A matrix that is the direct sum of Jordan blocks,

$$J = J_{k_1}(\lambda_1) \oplus J_{k_2}(\lambda_2) \oplus \cdots \oplus J_{k_t}(\lambda_t),$$

is called a *Jordan matrix.* In a Jordan matrix, t may equal 1 (that is, J may be a Jordan block), and the scalars $\lambda_1, \lambda_2, \ldots, \lambda_t$ need not be different. The characteristic polynomial of J equals

$$(\lambda - \lambda_1)^{k_1}(\lambda - \lambda_2)^{k_2} \cdots (\lambda - \lambda_t)^{k_t},$$

and hence its eigenvalues are

$$\lambda_1 \ (k_1 \text{ times}), \lambda_2 \ (k_2 \text{ times}), \ldots, \lambda_t \ (k_t \text{ times}),$$

the n scalars that occur on the main diagonal of J. The scalars $\lambda_1, \lambda_2, \ldots, \lambda_t$ are not necessarily distinct, so that the algebraic multiplicities of the eigenvalues are not necessarily $k_1, k_2, \ldots, k_t$. If μ is one of the numbers $\lambda_1, \lambda_2, \ldots, \lambda_t$, then the algebraic multiplicity of μ equals the sum of the orders of the Jordan blocks whose diagonal entries equal μ. Each Jordan block with diagonal entries equal to μ contributes 1 to the geometric multiplicity of μ, and hence the geometric multiplicity of μ equals the number of Jordan blocks containing μ on its main diagonal. Finally, we note that the digraph $\widehat{D}(J)$ is a collection of vertex-disjoint paths with $k_1, k_2, \ldots, k_t$ vertices, respectively. □

Example 7.3.10 The matrix

$$J = \left[\begin{array}{c|ccc|cc|cccc}
5 & & O & & & O & & & O & \\
\hline
 & 5 & 1 & 0 & & & & & & \\
O & 0 & 5 & 1 & & O & & & O & \\
 & 0 & 0 & 5 & & & & & & \\
\hline
O & & O & & 8 & 1 & & & O & \\
 & & & & 0 & 8 & & & & \\
\hline
 & & & & & & 8 & 1 & 0 & 0 \\
O & & O & & & O & 0 & 8 & 1 & 0 \\
 & & & & & & 0 & 0 & 8 & 1 \\
 & & & & & & 0 & 0 & 0 & 8
\end{array}\right]$$

is a Jordan matrix $J_1(5) \oplus J_3(5) \oplus J_2(8) \oplus J_4(8)$ of order 10. The eigenvalues of J are 5 with algebraic multiplicity $1 + 3 = 4$ and 8 with algebraic multiplicity $2 + 4 = 6$. The geometric multiplicity of 5 equals 2, the number of Jordan blocks with 5 on their main diagonal; the geometric multiplicity of 8 also equals 2, the number of Jordan blocks with 8 on the main diagonal. □

Our goal is to show that every matrix is similar to a Jordan matrix. By Theorem 7.3.7, a matrix A of order n is similar to an upper triangular matrix T where the eigenvalues of A occur on the main diagonal of T in any specified order. We now specify that equal eigenvalues occur consecutively on the main diagonal of T. Suppose that A has at least two different eigenvalues and let μ be the one that occurs in the initial positions of the main diagonal of T. Thus T has the form

$$T = \left[\begin{array}{c|c} T_1 & X \\ \hline O & U \end{array}\right],$$

where T_1 is an upper triangular matrix each of whose diagonal entries equals μ and U is an upper triangular matrix none of whose diagonal entries equals μ. We now show that we may take $X = O$, that is, T is similar to the matrix

$$\left[\begin{array}{c|c} T_1 & O \\ \hline O & U \end{array}\right], \tag{7.8}$$

by using elementary combination similarities. Consider rows i and j of T where row i intersects T_1 and row j intersects U, and the elementary similarity $T' = I(c \cdot j + i)TI(-c \cdot j + i)$. Let a be the (i, j) entry of T and let the (j, j) entry of T be θ. Then $\theta \neq \mu$, and the (i, j) entry of T' equals $a + c(\theta - \mu)$. Thus, by choosing $c = -a/(\theta - \mu)$, the (i, j) entry of T' equals 0. Moreover, since T_1 and U are upper triangular, T' differs from T only in those positions of row i in columns $j, j + 1, \ldots, n$ and those positions of column j in rows $1, 2, \ldots, i$. It now follows that by a sequence of elementary combination similarities, we may make each entry of X equal to 0 with no change in T_1 and U. We do this row by row starting from the last row of T_1 and working up to the first row, and making 0 each entry of the current row beginning with its first entry and working to the right to its last entry.

If the matrix U in (7.8) does not have a constant main diagonal, we repeat the above argument on U. Eventually, we obtain that our original matrix A is similar to a matrix of the form

$$T_1 \oplus T_2 \oplus \cdots \oplus T_p,$$

where each T_i is an upper triangular matrix with a constant main diagonal, and no two of these constants are equal. The reduction of T by similarity to a Jordan matrix is complete once each of the T_i have been reduced by similarity to a Jordan matrix.

Thus we may now assume that T is an upper triangular matrix of order m, each of whose main diagonal entries equals μ. The proof is by induction on m. If $m = 1$, then $T = [\mu]$ is a Jordan block of order 1. Now let $m = 2$. Then

$$T = \begin{bmatrix} \mu & a \\ 0 & \mu \end{bmatrix}$$

for some scalar a. If $a = 0$, then T is a direct sum of two Jordan blocks of order 1 and so is a Jordan matrix. Suppose that $a \neq 0$. By an elementary diagonal similarity (multiply row 1 by $1/a$ and column 1 by a) we obtain the Jordan matrix

$$\begin{bmatrix} \mu & 1 \\ 0 & \mu \end{bmatrix}.$$

Now assume that $m > 2$. The leading principal submatrix T' of T of order $m-1$ is upper triangular with all μ's on its main diagonal. By the inductive hypothesis, there is an invertible matrix Q of order $m-1$ such that the matrix $S' = Q^{-1}T'Q$ is a Jordan matrix. Let $P = Q \oplus I_1$, an invertible matrix of order m with inverse $Q^{-1} \oplus I_1$. Then

$$S = P^{-1}TP = \left[\begin{array}{ccc|c} & & & * \\ & S' & & \vdots \\ & & & * \\ \hline 0 & \cdots & 0 & \mu \end{array}\right]. \tag{7.9}$$

If the last column of S is all zeros apart from μ at its end, then since S' is a Jordan matrix, S is also a Jordan matrix, and we are done. We now assume that there is a nonzero entry in the last column of S above its last entry.

First suppose that there is an entry $h \neq 0$ in the last column that is in the same row as an off-diagonal 1 in one of the Jordan blocks of S'. In this case, there is an elementary combination similarity that replaces h with 0 and otherwise does not change S. For example, if

$$S = \left[\begin{array}{cc|cccc|c} \mu & 1 & & & & & * \\ 0 & \mu & & O & & & * \\ \hline & & \mu & 1 & 0 & 0 & * \\ & O & 0 & \mu & 1 & 0 & h \\ & & 0 & 0 & \mu & 1 & * \\ & & 0 & 0 & 0 & \mu & * \\ \hline 0 & 0 & 0 & 0 & 0 & 0 & \mu \end{array}\right], \tag{7.10}$$

the h is in the same row as the 1 in column 5. The elementary combination similarity that adds $-h$ times column 5 to column 7 and h times row 7 to row 5 replaces h with 0 and otherwise does not change S. In this way we can reduce the matrix S in (7.9) by elementary combination similarities so that the only nonzero entries in its last column above the μ in its last position occur in the same row as the last row of one of the Jordan blocks of S'. We now assume that S has this form. For instance, in the case S as given in (7.10), we get

$$S' = \left[\begin{array}{cc|cccc|c} \mu & 1 & & & & & 0 \\ 0 & \mu & & O & & & p \\ \hline & & \mu & 1 & 0 & 0 & 0 \\ & & 0 & \mu & 1 & 0 & 0 \\ & O & 0 & 0 & \mu & 1 & 0 \\ & & 0 & 0 & 0 & \mu & q \\ \hline 0 & 0 & 0 & 0 & 0 & 0 & \mu \end{array}\right], \tag{7.11}$$

where p and q may be nonzero.

The digraph $\widehat{D}(S)$ has a very simple structure. It consists of a number of pairwise vertex-disjoint paths (these correspond to the Jordan blocks that have only 0's across from them in column n) and, entirely disjoint from them, a number of other paths that, except for the fact that they all terminate at the vertex n (corresponding to column n), are also pairwise vertex-disjoint (these correspond to the Jordan blocks that have one nonzero entry across from their last row in column n). We now show that by elementary combination similarities we can replace all but one of the nonzero off-diagonal entries in column n of S with 0, again without changing any other entry of S. The nonzero off-diagonal entry that remains is one corresponding to the largest Jordan block of S' (if there is more than one such largest Jordan block, we can choose one arbitrarily). We refer to the particular S in (7.11), but our procedure works in general. The digraph $\widehat{D}(S)$ in this case consists of a path of length 2 and a path of length 5 that meet in the vertex corresponding to column 7. Assume that $p \neq 0$ and $q \neq 0$. The scalar q is opposite the largest Jordan block. Using an elementary diagonal similarity, we may assume that $q = 1$. With this in mind, we now perform a sequence of elementary similarities that replaces p with 0 and otherwise makes no change:

(i) Add $-p$ times row 6 to row 2 and p times column 2 to column 6.

(ii) Add $-p$ times row 5 to row 1 and p times column 1 to column 6.

Notice how this uses the fact that the second Jordan block (of order 4) has more rows that the first Jordan block (of order 2). In this way we reduce S to a Jordan matrix. Hence, by induction, we have proved the following important theorem.

Theorem 7.3.11 *Every matrix of order n is similar to a Jordan matrix.* □

If J is a Jordan matrix similar to the matrix A, then J is called a *Jordan Canonical Form* of A, abbreviated to JCF. A JCF of A is unique apart from the obvious fact that the Jordan blocks may occur in any order. For example,

$$J_2(5) \oplus J_3(5) \oplus J_2(8) \text{ and } J_3(5) \oplus J_2(8) \oplus J_2(5)$$

are both JCFs of the same matrix A. Indeed, such an A would have six JCFs, as there are $3! = 6$ ways in which to order the three different Jordan blocks. On the other hand, the matrix

$$J_3(5) \oplus J_3(5)$$

is the unique JCF of a matrix B as there is only one way to list its two (identical) Jordan blocks. We do not prove here the general uniqueness property of the JCF.

We have seen how a large part of the proof for a JCF—starting from Jacobi's theorem that a matrix is similar to a triangular matrix—can be made graph-theoretical (see [6] and the reference to Turnbull and Aitken there). A similar proof also appears in [23].

7.4 Spectrum of Circulants

We recall the definition of a circulant matrix from Section 3.2. Let P be the permutation matrix of order n defined by

$$P = \begin{bmatrix} 0 & 1 & 0 & \cdots & 0 & 0 \\ 0 & 0 & 1 & \cdots & 0 & 0 \\ 0 & 0 & 0 & \cdots & 0 & 0 \\ \vdots & \vdots & \vdots & \ddots & \vdots & \vdots \\ 0 & 0 & 0 & \cdots & 0 & 1 \\ 1 & 0 & 0 & \cdots & 0 & 0 \end{bmatrix}.$$

Let

$$g(x) = a_0 + a_1x + a_2x^2 + \cdots + a_{n-1}x^{n-1}$$

be a polynomial of degree at most n. Then

$$A = g(P) = a_0I_n + c_1P + c_2P^2 + \cdots + a_{n-1}P^{n-1}$$

is a circulant of order n.

The digraph $D(P)$ is a cycle of length n and thus P satisfies $P^n = I_n$. The characteristic polynomial of P is $\det(\lambda I_n - P)$. The digraph $D^*(\lambda I_n - P)$ is a cycle of length n with a loop at each of its vertices and hence has only two linear subdigraphs, namely, the cycle itself and the linear subdigraph consisting of the n loops. It follows from the definition of determinant that

$$det(\lambda I_n - P) = \lambda^n + (-1)^{n-1}(-1)^n = \lambda^n - 1.$$

Hence the eigenvalues of P are the n nth roots of unity $1, \omega, \omega^2, \ldots, \omega^{n-1}$, where $\omega = e^{2\pi i/n}$ and i is the complex number equal to the square root of -1. Because the eigenvalues of P are distinct, the Jordan Canonical Form of P has only Jordan blocks of order 1. Hence the Jordan Canonical Form of P is the diagonal matrix

$$D = \begin{bmatrix} 1 & 0 & 0 & \cdots & 0 \\ 0 & \omega & 0 & \cdots & 0 \\ 0 & 0 & \omega^2 & \cdots & 0 \\ \vdots & \vdots & \vdots & \ddots & 0 \\ 0 & 0 & 0 & \cdots & \omega^{n-1} \end{bmatrix}.$$

We now invoke Corollary 7.3.8 and conclude that the n eigenvalues of the circulant C are

$$g(\omega^k) \quad (k = 0, 1, \ldots, n-1).$$

Let

$$X = \begin{bmatrix} 1 & 1 & 1 & \cdots & 1 \\ 1 & \omega & \omega^2 & \cdots & \omega^{n-1} \\ 1 & \omega^2 & \omega^4 & \cdots & \omega^{2(n-1)} \\ \vdots & \vdots & \vdots & \ddots & \vdots \\ 1 & \omega^{n-1} & \omega^{2(n-1)} & \cdots & \omega^{(n-1)(n-1)} \end{bmatrix}.$$

Then columns of X are the eigenvectors for the n eigenvalues $1, \omega, \omega^2, \ldots, \omega^{n-1}$ of P and hence

$$PX = XD \text{ or, equivalently, } X^{-1}PX = D.$$

Since $C = g(P)$ we also have that

$$X^{-1}CX = g(D).$$

Hence the columns of X are also eigenvectors for the eigenvalues

$$g(1), g(\omega), g(\omega^2), \ldots, g(\omega^{n-1})$$

of C.

7.5 Exercises

1. Let x and y be eigenvectors for the eigenvalue λ of a square matrix A, and let α and β be real numbers. Prove that $\alpha x + \beta y$ is also an eigenvector for the eigenvalue λ of A.

2. Show that the eigenvalues of the matrix $aI_n + bJ_n$ are a ($n-1$ times) and $a + nb$ (once). Here J_n is the square matrix of order n, each of whose entries is 1.

3. A square matrix is *idempotent* provided $A^2 = A$. For example, the matrix
$$\begin{bmatrix} 1 & 1 \\ 0 & 0 \end{bmatrix}$$
is idempotent. Prove that 0 and 1 are the only possible eigenvalues of an idempotent matrix. (Note that the zero matrix is idempotent with each of its eigenvalues equal to 0, and the identity matrix is idempotent with each of its eigenvalues equal to 1.)

4. Prove that the trace and rank of an idempotent matrix are equal.

5. Let A and B be matrices of order n. Prove that λ is an eigenvalue of AB if and only if λ is an eigenvalue of BA.

6. Let A and B be matrices of order n, at least one of which is invertible. Show that AB and BA are similar.

7. Let A be a nonsingular matrix of order n. Determine the characteristic polynomial of A^{-1} in terms of the characteristic polynomial of A.

8. Let A be an invertible matrix of order n. Use the Cayley-Hamilton theorem to obtain a polynomial $f(x)$ such that $A^{-1} = f(A)$.

9. Determine the characteristic polynomial of a general matrix of order 4 by means of the Coates digraph.

10. Let u and v be vectors in $\Re^n$. Let A be the matrix uv^T of order n (the (i,j) of this matrix is $u_i v_j$ $(1 \leq i, j \leq n)$). Find the eigenvalues and eigenvectors of A.

11. Calculate the n eigenvalues of the matrix

$$\begin{bmatrix} 0 & 0 & \cdots & 0 & a_1 \\ 0 & 0 & \cdots & 0 & a_2 \\ \vdots & \vdots & \ddots & \vdots & \vdots \\ 0 & 0 & \cdots & 0 & a_{n-1} \\ a_1 & a_2 & \cdots & a_{n-1} & a_n \end{bmatrix}.$$

12. Determine all possible Jordan Canonical Forms for a matrix of order 6, all of whose eigenvalues equal 3. Classify these Jordan Canonical Forms according to the geometric multiplicity of 3.

13. Find the Jordan Canonical Form of the matrix

$$\begin{bmatrix} 0 & 1 & 0 & 0 \\ 0 & 0 & 1 & 0 \\ 0 & 0 & 0 & 1 \\ 1 & -4 & 6 & 4 \end{bmatrix}.$$

14. Determine the Jordan Canonical Form of the matrix

$$\begin{bmatrix} 2 & 0 & 0 & 0 \\ 0 & 1 & 1 & 1 \\ 0 & 0 & 2 & 0 \\ 1 & 0 & 0 & 1 \end{bmatrix}.$$

15. Let A be a matrix of order m and let B be a matrix of order n. Let $p_A(\lambda)$ be the characteristic polynomial of A. Prove that $p_A(B)$ is invertible if and only if A and B have no common eigenvalues.

16. Determine the eigenvalues of the matrix

$$\begin{bmatrix} 1 & 0 & -1 & 0 \\ 0 & 1 & 0 & -1 \\ -1 & 0 & 1 & 0 \\ 0 & -1 & 0 & 1 \end{bmatrix}.$$

17. For the Jordan block $J_k(0)$, show that $J_k(0)^k = O$ but $J_k(0)^{k-1} \neq O$. Let $p_k(x) = (x-a)^k$. Deduce that the Jordan block $J_k(a)$ satisfies $p_k(J_k(a)) = O$ but $p_{k-1}(J_k(a)) \neq O$.

18. Let A be a matrix of order n. Let $\lambda_1, \lambda_2, \ldots, \lambda_l$ be the distinct eigenvalues of A, and in the Jordan Canonical Form of A, let the largest Jordan block corresponding to the eigenvalues $\lambda_1, \lambda_2, \ldots, \lambda_l$ be, respectively, $m_1, m_2, \ldots, m_l$. Let

$$q(x) = (x - \lambda_1)^{m_1}(x - \lambda_2)^{m_2} \cdots (x - \lambda_l)^{m_l}.$$

Prove that $q(A) = O$ and that if any of the exponents $m_1, m_2, \ldots, m_l$ in $q(x)$ is decreased, resulting in the polynomial $p(x)$, then $p(A) \neq O$.

Chapter 8

Nonnegative Matrices

In this chapter we consider matrices each of whose entries is a nonnegative number. These matrices have special spectral properties that depend solely on the digraph of the matrix and are independent of the magnitude of the positive entries. Some important classes of nonnegative matrices, such as irreducible (Section 8.1), primitive, and imprimitive matrices (Section 8.2), are defined here, contrary to the standard approach, by properties of associated digraphs (strong connectednes, lengths of cycles etc.). We discuss, mostly without proof, many of the results of the so-called Perron–Frobenius theory of nonnegative matrices (Section 8.3). This theory represents a basic ingredient of the theory of graph spectra where tools from matrix theory are used to study graphs (a direction quite opposite from our main stream here since we want to show how graphs are used to treat matrices). Section 8.4 represents a short introduction to graph spectra.

8.1 Irreducible and Reducible Matrices

A matrix is called *nonnegative*, respectively, *positive*, provided all of its entries are nonnegative, respectively, positive. In the theory of nonnegative matrices the notion of irreducibility plays an important role, and this is equivalent to the notion of strong connectivity for digraphs discussed in Chapter 1.

Definition 8.1.1 A square matrix A of order n is *irreducible* provided that its digraph $D(A)$ is strongly connected; otherwise, A is *reducible.* Recall from Theorem 1.2.3 that a digraph is strongly connected if and only if there does not exist a partition of its vertex set into two nonempty sets U and W such that each edge between U and W has itsr initial vertex in U and its terminal vertex in W. Thus if we simultaneously permute the rows and columns of A so that the first rows and columns correspond to U, we obtain that A is reducible if and only if there is a permutation matrix P such that

$$PAP^T = \begin{bmatrix} X & Y \\ O & Z \end{bmatrix}, \tag{8.1}$$

where X and Z are square matrices of order at least 1. The matrix A is irreducible provided the form (8.1) cannot be achieved for any permutation matrix P. Note that the zero matrix in (8.1) is of size p by q where p and q are positive integers with $p + q = n$. It follows from the definition that a matrix of order 1 is always irreducible. We also note that if we had listed the vertices of W first, then we would get a permutation matrix Q such that

$$QAQ^T = \begin{bmatrix} Z & O \\ Y & X \end{bmatrix},$$

with the zero matrix occuring in the lower right corner. □

Recall from Theorem 1.2.3 that a digraph G has $l \geq 1$ strong components (strongly connected, induced subdigraphs whose sets of vertices partition the set of vertices of G) and that these strong components can be ordered as $G_1, G_2, \ldots, G_l$ so that the only edges between the components are edges whose initial vertex is a vertex in G_i and whose terminal vertex is a vertex in G_j, where $i < j$, that is, in the ordering $G_1, G_2, \ldots, G_l$, all edges between components go from left to right. We have $l = 1$ if and only if G is strongly connected. Applying this fact to the digraph $D(A)$ of a square matrix A, we see that the rows and columns of A can be simultaneously permuted to achieve a block diagonal form called the *Frobenius normal form:*

There exists a permutation matrix Q such that

$$QAQ^T = \begin{bmatrix} A_1 & A_{12} & A_{13} & \cdots & A_{1l} \\ O & A_2 & A_{23} & \cdots & A_{2l} \\ O & O & A_3 & \cdots & A_{3l} \\ \vdots & \vdots & \vdots & \ddots & \vdots \\ O & O & O & \cdots & A_l \end{bmatrix}, \tag{8.2}$$

where $A_1, A_2, \ldots, A_l$ are irreducible square matrices. Because the matrices $A_1, A_2, \ldots, A_l$ correspond to the strong components of $D(A)$, and since strong components are uniquely determined (they are the equivalence classes of an equivalence relation on the vertices of $D(A)$), the matrices $A_1, A_2, \ldots, A_l$ are uniquely determined up to simultaneous permutations of their rows and columns, that is, up to the order in which the vertices of the strong components are written down. The matrices $A_1, A_2, \ldots, A_l$ are the *irreducible components* of A. The matrix A is irreducible if and only if it has exactly one irreducible component. The order in which the irreducible components occur on the diagonal in (8.1) is not necessarily unique; it all depends on whether or not the matrices A_{ij} are zero matrices.

Example 8.1.2 The following matrix is in Frobenius normal form:

$$\left[\begin{array}{cc|cc|cc|cc} 0 & 1 & 0 & 0 & 0 & 0 & 0 & 0 \\ 1 & 1 & 0 & 0 & 0 & 0 & 0 & 0 \\ \hline 0 & 0 & 0 & 1 & 1 & 0 & 0 & 1 \\ 0 & 0 & 1 & 0 & 0 & 1 & 0 & 1 \\ \hline 0 & 0 & 0 & 0 & 1 & 1 & 1 & 1 \\ 0 & 0 & 0 & 0 & 1 & 0 & 0 & 0 \\ \hline 0 & 0 & 0 & 0 & 0 & 0 & 1 & 1 \\ 0 & 0 & 0 & 0 & 0 & 0 & 1 & 1 \end{array}\right].$$

There are four irreducible components, and the first irreducible component could be in any one of the four places; the relative order of the other three irreducible components is fixed. □

An important algebraic characterization of irreducible, nonnegative matrices is contained in the following theorem.

Theorem 8.1.3 *Let A be a nonnegative matrix of order n. Then A is irreducible if and only if $(I + A)^{n-1}$ is a positive matrix, equivalently, $I_n + A + A^2 + \cdots + A^{n-1}$ is a positive matrix.*

Proof. The matrix $I_n + A$ has positive diagonal entries and hence the digraph $D(I_n + A)$ has a loop at each vertex. First suppose that A is irreducible. Then $D(I_n + A)$ is strongly connected and for each ordered pair u, v of distinct vertices there is a (shortest) path from u to v of length at most $n - 1$. Because there is a loop at each vertex, there is a walk of length exactly $n - 1$ from u to v. Because $I_n + A$ is a nonnegative matrix, all walks have positive weights. It follows that $(I_n + A)^{n-1}$ is a positive matrix. Since

$$(I_n + A)^{n-1} = I_n + \binom{n}{1} A + \binom{n}{2} A^2 + \cdots + \binom{n-1}{n-1} A^{n-1},$$

it also follows that $I_n + A + A^2 + \cdots + A^{n-1}$ is a positive matrix.

Conversely, if $(I + A)^{n-1}$ is a positive matrix, then for each ordered pair of distinct vertices u, v there is a path of positive weight in $D(A)$ from u to v of length $n-1$. Hence $D(A)$ is strongly connected and A is irreducible. □

The following corollary is an easy consequence of Theorem 8.1.3.

Corollary 8.1.4 *Let A be an irreducible nonnegative matrix of order n each of whose diagonal entries is positive. Then A^{n-1} is a positive matrix.* □

8.2 Primitive and Imprimitive Matrices

The cycles of the (strongly connected) digraph $D(A)$ of an irreducible nonnegative matrix A have a strong influence on the spectrum of A.

Definition 8.2.1 Let G be a strongly connected digraph. The greatest common divisor (abbreviated GCD) d of the lengths of the cycles of G is called the *index of imprimitivity* of G. If $d = 1$, then G is *primitive*; if $d > 1$, then G is *imprimitive*. Note that since a closed walk is composed of cycles, in defining d we could use the lengths of all the closed walks in G (see also Lemma 8.2.2 below). Let A be an irreducible nonnegative matrix of order n. Then the *index of imprimitivity* of A is defined to be the index of imprimitivity of the digraph $D(A)$, and A is *primitive* or *imprimitive* according to whether $D(A)$ is primitive or imprimitive. □

If the index of imprimitivity d is greater than 1, a certain structure is imposed on a digraph and a matrix. First we make the following observation.

Lemma 8.2.2 *Let G be a strongly connected digraph with vertices $1, 2, \ldots, n$ and with index of imprimitivity d. For each vertex i of G, let d_i be the GCD of the lengths of all closed walks containing i. Then $d = d_1 = d_2 = \cdots = d_n$. Moreover, the lengths of any two walks with the same initial vertex and the same terminal vertex are congruent modulo d.*

Proof. Consider vertices i and j with $i \neq j$. Because G is strongly connected, there exists a path γ from i to j and a path γ' from j to i. The path γ followed by γ' gives a closed walk θ containing both i and j. Let θ have length s. Then d_i and d_j are both divisors of s. Thus, for each integer l for which there exists a closed walk of length l containing vertex i, there exists a closed walk of length $s + l$ containing vertex j. Because d_j is a divisor of s and of $s + l$, d_j is a divisor of l. Because this is true for all such l, we conclude that d_j is a divisor of d_i. In a similar way one shows that d_i is a divisor of d_j. Thus $d_i = d_j$ and we conclude that $d_1 = d_2 = \cdots = d_n$. The common value must be d.

Now let γ_1 and γ_2 be two walks with the same initial vertex i and the same terminal vertex j of lengths k_1 and k_2, respectively. There exists a walk γ_3 from vertex j to vertex i of some length t, giving two closed walks of lengths $k_1 + t$ and $k_2 + t$. Because d

divides both k_1+t and k_2+t, d divides $(k_1+t)-(k_2+t)=k_1-k_2$, that is, k_1 and k_2 are congruent modulo d. □

Theorem 8.2.3 *Let G be a strongly connected digraph with vertex set V having an index of imprimitivity equal to $d>1$. Then V can be partitioned into d nonempty sets $V_0, V_1, \ldots, V_{d-1}$ such that each edge of G has its initial vertex in some V_i and its terminal vertex in V_{i+1} (subscripts considered modulo d). Thus the subdigraphs induced on each of $V_0, V_1, \ldots, V_{d-1}$ do not contain any edges, and the edges of G are arranged in a circular pattern V_0 to V_1, V_1 to V_2, ... , V_{d-2} to V_{d-1}, and V_{d-1} to V_0.*

Proof. Consider any vertex a of G. For $i=0,1,\ldots,d-1$, let V_i be the set of vertices x to which there is some walk from a of length congruent to i modulo d (and so by Lemma 8.2.2, every walk from a to x has length congruent to i modulo d). Note that a belongs to V_0, and the sets $V_0, V_1, \ldots, V_{d-1}$ are pairwise disjoint and nonempty (because there is a cycle containing a and it has length at least d). Suppose that there is an arc from vertex u to vertex v, where u is in V_i and v is in V_j. There is a walk from a to u of length congruent to i modulo d and hence a walk from a to v of length congruent to $i+1$ modulo d. From the definition of the $V_k's$ we now conclude that $i+1$ is congruent to j modulo d, that is, modulo d, j equals $i+1$. □

The pairwise disjoint sets $V_0, V_1, \ldots, V_{d-1}$ in Theorem 8.2.3 are called the *sets of imprimitivity* of G. In case $d=1$, the vertex set V is the unique set of imprimitivity of G. A digraph G with the structure as given in Theorem 8.2.3 is called *cyclically d-partite.* Note that if m is a divisor of d, then G is also cyclically m-partite.

If G is the digraph of an irreducible nonnegative matrix and we order the vertices of G as given in Theorem 8.2.3, so that the vertices in V_0 come first, followed by those in V_2, ... , followed by those in V_{d-1}, we obtain the the following matrix interpretation of Theorem 8.2.3.

Theorem 8.2.4 *Let A be an irreducible nonnegative matrix of order n with index of imprimitivity equal to d. Then there exists a permutation matrix P such that*

$$PAP^T = \begin{bmatrix} O_{k_0} & A_{01} & O & \cdots & O & O \\ O & O_{k_1} & A_{12} & \cdots & O & O \\ O & O & O_{k_2} & \cdots & O & O \\ \vdots & \vdots & \vdots & \ddots & \vdots & \vdots \\ O & O & O & \cdots & O_{k_{d-2}} & A_{d-2,d-1} \\ A_{d-1,0} & O & O & \cdots & O & O_{k_{d-1}} \end{bmatrix}, \tag{8.3}$$

where the square zero matrices on the main diagonal have orders $k_0, k_1, \ldots, k_{d-1}$ as indicated (these are the sizes of $V_0, V_1, \ldots, V_{d-1}$ in Theorem 8.2.3). □

In general, the matrices $A_{i,i+1}$ in (8.3) are rectangular. Using block multiplication, we immediately obtain the following corollary.

Corollary 8.2.5 *Let A be an irreducible nonnegative matrix of order n with index of imprimitivity equal to d. Then there exists a permutation matrix P such that*

$$PA^dP^T = B_0 \oplus B_1 \oplus \cdots \oplus B_{d-1},$$

a block-diagonal matrix whose blocks $B_0, B_1, \ldots, B_{d-1}$ are

$$B_0 = A_{01}A_{12}\cdots A_{d-1,0},\ B_1 = A_{12}A_{23}\cdots A_{01}, \ldots,$$

$$B_{d-1} = A_{d-1,0}A_{01}\cdots A_{d-2,d-1}.$$

□

To conclude this section, we show that some positive integral power of a primitive (nonnegative) matrix is a positive matrix. In order to do this, we make use of a number-theoretic lemma that we state without proof.

Lemma 8.2.6 *Let $d_1, d_2, \ldots, d_k$ be positive integers whose GCD equals 1. Then every sufficiently large positive integer can be expressed as a nonnegative linear combination of $d_1, d_2, \ldots, d_k$. That is, there exists a positive integer M such that, for each integer $m \geq M$, there are nonnegative integers $a_1, a_2, \ldots, a_k$ such that*

$$m = a_1 d_1 + a_2 d_2 + \cdots + a_k d_k.$$

Theorem 8.2.7 *Let A be a primitive matrix of order n. Then there exists a positive integer p such that A^p is a positive matrix.*

Proof. The matrix A^p is a positive matrix if and only if, in the digraph $D(A)$, for any ordered pair of not necessarily distinct vertices u, v there is a walk of length p from u to v. Because A is primitive, the digraph $D(A)$ is strongly connected and the GCD of the lengths of its cycles equals 1. Let the distinct cycle lengths of $D(A)$ be $d_1, d_2, \ldots, d_k$, where $1 \leq d_1, d_2, \ldots, d_k \leq n$. Because $D(A)$ is strongly connected, we can find a walk γ_{uv} from u to v that contains a vertex of a cycle of each length $d_1, d_2, \ldots, d_k$. Let the length of such a walk be l_{uv}. We can extend γ_{uv} by going around the cycles it meets any number of times. Applying Lemma 8.2.6, we can obtain walks from u to v of any length greater than or equal to $l_{uv} + M$. Now let p be the maximum of the numbers $l_{uv} + M$ taken over all ordered pairs of vertices u, v. Then there is a walk of length p from u to v for all u and v. Hence A^p is a positive matrix. $\square$

If A is a primitive matrix of order n, then the smallest positive power of A that gives a positive matrix is called the *exponent* of A. It is known that the exponent of A is at most $n^2 - 2n + 2$. This exponent is achieved by the following matrix of order n with $n+1$ positive entries:

$$\begin{bmatrix} 0 & a_1 & b & 0 & \cdots & 0 \\ 0 & 0 & a_2 & 0 & \cdots & 0 \\ 0 & 0 & 0 & a_3 & \cdots & 0 \\ \vdots & \vdots & \vdots & \vdots & \ddots & \vdots \\ 0 & 0 & 0 & 0 & \cdots & a_{n-1} \\ a_n & 0 & 0 & 0 & \cdots & 0 \end{bmatrix},$$

where $a_1, a_2, \ldots, a_n, b$ are all positive. Note that because clearly a primitive matrix (or irreducible matrix of order at least 2) has at least one positive entry per row and column, it follows that if A^p is a positive matrix, so are all powers of A greater than p. A power of an imprimitive matrix cannot be positive. This follows, for instance, from Lemma 8.2.2. Thus we can say that *some power of a nonnegative square matrix is a positive matrix if and only if the matrix is primitive.*

8.3 The Perron–Frobenius Theorem

The spectra of irreducible nonnegative matrices, in particular of positive matrices, have many special properties, which we present in this section without proof. First we make a general definition.

Definition 8.3.1 Let A be a matrix of order n with eigenvalues $\lambda_1, \lambda_2, \ldots, \lambda_n$. The *spectral radius* $\rho(A)$ of A is the maximum of the absolute values of its eigenvalues:

$$\rho(A) = \max\{|\lambda_1|, |\lambda_2|, \ldots, , |\lambda_n|\}.$$

The spectral radius of A is the radius of the smallest circle centered at the origin that contains the spectrum of A. This circle is called the *spectral circle* of A. The spectral radius of A is zero if A is a nilpotent matrix and is positive otherwise.

Let A be an irreducible nonnegative matrix of order n. If $n > 1$, then the digraph $D(A)$ has a closed walk and hence cannot be nilpotent; hence A has positive spectral radius. □

Example 8.3.2 Let

$$A = \begin{bmatrix} 0 & 1 & 0 \\ 0 & 0 & 1 \\ 0 & 0 & 0 \end{bmatrix},$$

a nonnegative reducible matrix each of whose three irreducible components is the zero matrix of order 1. Then $A^3 = O$, and the eigenvalues of A are $0, 0, 0$. Hence the spectral radius of A is 0. Note that A is the Jordan block $J_3(0)$. □

The following theorem gives an elementary bound for the spectral radius of a matrix.

Theorem 8.3.3 *Let $A = [a_{ij}]$ be a matrix of order n. Let*

$$r_i = \sum_{j=1}^{n} |a_{ij}| \quad (1 \le i \le n)$$

be the sum of the absolute values of the entries in row i of A. Then

$$\rho(A) \le \max\{r_1, r_2, \ldots, r_n\}.$$

A similar inequality holds for the sum of the absolute values of the entries in the columns of A.

Proof. Let λ be any eigenvalue of A and let $x = [x_1\ x_2\ \ldots\ x_n]^T$ be a corresponding eigenvector: $Ax = \lambda x$. Let $|x_k| = \max\{|x_i| : 1 \le i \le n\} > 0$ be the largest absolute value of an entry of x. Then, taking absolute values in the equation

$$\sum_{j=1}^{n} a_{kj}x_j = \lambda x_k$$

and using the triangle inequality, we obtain

$$\begin{aligned} |\lambda||x_k| = |\lambda x_k| &= \left|\sum_{j=1}^{n} a_{kj}x_j\right| \\ &\le \sum_{j=1}^{n} |a_{kj}||x_j| \\ &\le (\sum_{j=1}^{n} |a_{kj}|)|x_k| = r_k|x_k|. \end{aligned}$$

Cancelling $|x_k|$, we obtain $|\lambda| \le r_k \le \max\{r_1, r_2, \ldots, r_n\}$. □

In the next theorem we summarize the most important and basic consequences of the Perron–Frobenius theory of nonnegative matrices. In order to avoid the trivial situation of a zero matrix of order 1 (which is an irreducible nonnegative matrix), we assume that $n > 1$.

Theorem 8.3.4 *Let A be an irreducible nonnegative matrix of order $n > 1$. Then*

(i) *The spectral radius $\rho(A)$ of A is an eigenvalue of A, that is, A has a positive eigenvalue r that is greater than or equal to the absolute value of every eigenvalue of A. The number r, which is the same as the spectral radius of A, is sometimes called the Perron eigenvalue of A.*

(ii) *The algebraic multiplicity, and so the geometric multiplicity, of the Perron eigenvalue r equals 1, that is, r is a simple root of the characteristic polynomial of A.*

(iii) *Corresponding to the Perron eigenvalue r there is a positive eigenvector y: $Ay = ry$, where y is a positive vector. The vector y, and each of its positive multiples, is called a Perron vector of A. The matrix A has no other nonnegative eigenvectors (corresponding to any eigenvalue) other than positive multiples of its Perron vector.*

(iv) *Let h be the index of imprimitivity of A. Then A has exactly h eigenvalues whose absolute value equals r, that is, there are exactly h eigenvalues on the spectral circle of A. The h eigenvalues of A on the spectral circle are the roots of the equation $\lambda^h - r^h = 0$, that is, the numbers*

$$re^{2\pi i \cdot j/h} \quad (j = 0, 1, \ldots, h-1).$$

In fact, the entire spectrum of A is mapped into itself under a rotation of the plane about the origin through an angle of $2\pi/h$.

(v) *If A' is a principal submatrix of A, then $\rho(A') \leq \rho(A)$ with equality if and only if $A' = A$.*

(vi) *If B is a nonnegative matrix with $B \leq A$ (entrywise), then $\rho(B) \leq \rho(A)$ with equality if and only if $B = A$.*

Example 8.3.5 Let P_n be the permutation matrix whose digraph $D(P)$ has n edges arranged in the cycle that goes from 1 to 2, from 2 to 3, ... , from $n-1$ to n, and from n to 1. For instance,

$$P_4 = \begin{bmatrix} 0 & 1 & 0 & 0 \\ 0 & 0 & 1 & 0 \\ 0 & 0 & 0 & 1 \\ 1 & 0 & 0 & 0 \end{bmatrix}.$$

The matrix P_n is an irreducible nonnegative matrix with spectral radius equal to 1 and with index of imprimitivity equal to n. Its n eigenvalues are

$$e^{2\pi i \cdot j/n} \quad (j = 0, 1, \ldots, n-1).$$

When $j = 0$, we get the Perron eigenvalue 1. A Perron eigenvector is the vector of all 1's or, more generally, a constant vector each of whose entries is a positive number c.

Now let

$$A = \begin{bmatrix} 0 & 0 & 0 & 1 & 1 & 1 \\ 0 & 0 & 0 & 1 & 1 & 1 \\ 0 & 0 & 0 & 1 & 1 & 1 \\ 1 & 1 & 1 & 0 & 0 & 0 \\ 1 & 1 & 1 & 0 & 0 & 0 \\ 1 & 1 & 1 & 0 & 0 & 0 \end{bmatrix}.$$

The matrix A is the adjacency matrix of the complete bipartite graph $K_{3,3}$. Then A is irreducible and has an index of imprimitivity equal to 2, and $A^2 = 3J_3 \oplus 3J_3$. The matrix J_3 has eigenvalues $3, 0, 0$ and hence A^2 has eigenvalues $9, 9, 0, 0, 0, 0$. The eigenvalues of A are $3, -3, 0, 0, 0, 0$ (because A has trace equal to zero, the sum of the eigenvalues equals 0). Hence the Perron eigenvalue of A equals 3 and a Perron vector is $[1\ 1\ 1\ 1\ 1\ 1]^T$.

□

By Theorem 8.3.3, if A is an irreducible nonnegative matrix, then the maximum sum of the elements in a row of A is an upper bound for the spectral radius of A. Theorem 8.3.4 enables us to obtain a lower bound as well.

Theorem 8.3.6 *Let $A = [a_{ij}]$ be an irreducible nonnegative matrix of order $n > 1$. Let*

$$r_i = \sum_{j=1}^{n} a_{ij} \quad (1 \le i \le n).$$

Then

$$\min\{r_i : 1 \le i \le n\} \le \rho(A) \le \max\{r_i : 1 \le i \le n\}.$$

Equality occurs on the left if and only if it occurs on the right, and this happens if and only if $r_1 = r_2 = \cdots = r_n$.

Proof. Let $y = [y_1\ y_2\ \cdots\ y_n]^T$ be a Perron vector corresponding to the Perron eigenvalue $\rho(A)$. Then y is a positive vector. Let

$$y_s = \min\{y_1, y_2, \ldots, y_n\} \text{ and } y_t = \max\{y_1, y_2, \ldots, y_n\}.$$

Then $y_s, y_t > 0$, and from the equations

$$\sum_{j=1}^{n} a_{sj} y_j = \rho(A) y_s \text{ and } \sum_{j=1}^{n} a_{tj} y_j = \rho(A) y_t$$

we get, similar to the proof of Theorem 8.3.3, that

$$y_s r_s \le \rho(A) y_s \text{ and } \rho(A) y_t \le y_t r_t,$$

and hence

$$\min\{r_i : 1 < i \le n\} \le \rho(A) \le \max\{r_i : 1 \le i \le n\}. \qquad (8.4)$$

It is straightforward to check that equality holds in either of the two inequalities in (8.4) if and only if the Perron eigenvector y is a constant vector. But a constant vector is a Perron eigenvector if and only if $r_1 = r_2 = \cdots = r_n$, and the theorem now follows. □

As we have seen, the Perron–Frobenius theory of nonnegative matrices depends substantially on the zero-nonzero pattern of a matrix, and this translates to the digraph. For instance, an irreducible matrix becomes a strongly connected digraph; for more on this, one may consult [3] and [7]. We conclude this brief introduction to spectral properties of nonnegative matrices by mentioning some applications to the adjacency matrices of graphs (more generally, multigraphs). (The Perron–Frobenius theorem can also be applied to the adjacency matrix of a multidigraph, but we shall not go in this direction.)

8.4 Graph Spectra

In the theory of graph spectra (see, for example, [18], [71], [24]), the results of matrix theory are used for investigations of graphs. In this section we present some basic properties of graph spectra. In Sections 10.2 and 10.3 we discuss some applications of graph spectra in physics and chemistry that also provide motivation for founding the theory.

We start with the following definition:

Definition 8.4.1 Let G be a multigraph whose vertex set is $\{1, 2, \ldots, n\}$, and let $A = [a_{ij}]$ be an adjacency matrix of G. Then a_{ij} equals the number of edges between vertices i and j and hence A is a nonnegative, symmetric integral matrix of order n. By Theorem 7.1.9. because A is a real symmetric matrix, each of its eigenvalues $\lambda_1, \lambda_2, \ldots, \lambda_n$ is real, and we may choose our notation so that

$$\lambda_1 \geq \lambda_2 \geq \cdots \geq \lambda_n.$$

The characteristic polynomial of A is called the *characteristic polynomial of the multigraph G*, and the eigenvalues of A are called the *eigenvalues of the multigraph G*. The *spectrum of the multigraph G* is the collection of its n eigenvalues $\lambda_1, \lambda_2, \ldots, \lambda_n$. By Theorem 8.3.4, the eigenvalue $r = \lambda_1$ is the spectral radius of A and $\lambda_n \geq -r$. Thus the spectrum of G lies in the interval $[-r, r]$, where r is the largest eigenvalue. The eigenvalue r is called the *index of the multigraph G*.

We now restrict our attention to graphs G. Thus G has no loops and at most one edge joins each pair of vertices. The adjacency matrix A of G is a symmetric matrix of 0's and 1's with only 0's on the main diagonal.

Example 8.4.2 The complete graph K_n has adjacency matrix $A = J_n - I_n$. All the row and column sums of A equal $n-1$ and so, by Theorem 8.3.6, the index of K_n equals $n-1$. We have $(-1)I_n - A = -J_n$. Because $-J_n$ has rank 1, this implies that -1 is an eigenvalue of A with geometric, and thus algebraic, multiplicity equal to $n-1$. Thus the eigenvalues of K_n are $n-1, 0, \ldots, 0$ ($n-1$ zeros).

Now let G be the complete bipartite graph $K_{p,q}$. Then an adjacency matrix is

$$A = \begin{bmatrix} O_p & J_{p,q} \\ J_{q,p} & O_q \end{bmatrix}.$$

Squaring A we see that

$$A^2 = \begin{bmatrix} qJ_p & O \\ O & pJ_q \end{bmatrix}.$$

The eigenvalues of A^2 are

$$pq, \underbrace{0, \ldots, 0}_{p-1}, pq, \underbrace{0, \ldots, 0}_{q-1}$$

and, because the trace of A equals 0, the eigenvalues of $K_{p,q}$ are $\pm\sqrt{pq}$ followed by $(p+q-2)$ 0's.

If G is a connected bipartite graph with index r, then all cycles have even length and hence the index of imprimitivity of its adjacency matrix (regarded as an adjacency matrix of a digraph and so with an edge from a vertex u to a vertex v if and only if there is an edge from vertex v to u as well) is a multiple of 2; hence it follows from (iv) of Theorem 8.3.4 that the spectrum is symmetric about zero, in particular, $-r$ is also an eigenvalue of G. If G is not bipartite, then the index of imprimitivity of G is 1, and hence $-r$ cannot be an eigenvalue of G.

Theorem 8.4.3 *Let G be a graph whose eigenvalues are $\lambda_1 \geq \lambda_2 \geq \ldots \geq \lambda_n$, let $r = \lambda_1$ be the index of G, and let $s = \lambda$ be the smallest eigenvalue of G. Then*

(i) *The eigenvalues* $\lambda_1, \lambda_2, \ldots, \lambda_n$ *are real and satisfy*

$$\lambda_1 + \lambda_2 + \cdots + \lambda_n = 0.$$

(ii) *The number of edges of* G *equals*

$$\frac{\lambda_1^2 + \lambda_2^2 + \cdots + \lambda_n^2}{2}.$$

(iii) *If* G *has no edges, then all if its eigenvalues equal* 0.

(iv) *If* G *has at least one edge, then* $1 \leq r \leq n-1$ *and* $-r \leq s \leq -1$. *We have* $r = n-1$ *if and only if* $G = K_n$ *and* $r = 1$ *if and only if each connected component of* G *is either* K_1 *or* K_2 *(there must be at least one* K_2 *because* G *has at least one edge). We also have* $s = -1$ *if and only if each connected component of* G *is a complete graph, and* $s = -r$ *if and only if the connected component of* G *with the largest index is a bipartie graph.*

Proof. Let A be the adjacency matrix of G. Then A is symmetric with trace equal to zero. Hence (i) holds. Then the number of edges of G equals the number of closed walks of length 2, and this equals the trace of A^2 divided by 2 (because each edge has two vertices). Because A^2 has eigenvalues $\lambda_1^2, \lambda_2^2, \ldots, \lambda_n^2$, (ii) holds. The adjacency matrix of a graph with no edges is a zero matrix, and (iii) follows. Because the index of K_n is $n-1$, every subgraph of K_n not equal to K_n has a strictly smaller index by (v) and (vi) of Theorem 8.3.4. Because G has at least one edge, K_2 is a subgraph of G where the spectrum of K_2 is $1, -1$. The assertions in (iv) about r now follow easily. Because K_2 has least eigenvalue equal to -1, it follows from the interlacing theorem for symmetric matrices [47] that $q \leq -1$ with equality if and only if each connected component is a complete graph. That $q \geq -r$ follows since the index of G is at most r and is the largest eigenvalue in absolute value. That $q = -r$ if and only if the connected component of largest index is bipartite follows from the last assertion in Example 8.4.2. □

When considering determinants of symmetric matrices, in particular adjacency matrices of graphs, it is useful to introduce a concept that is related to linear subdigraphs.

Let G be a graph. A subgraph of G whose components are circuits or graphs K_2 is called a *basic figure* of G. A basic figure is a *spanning basic figure* if it contains all vertices of G. If U is a basic figure, then $p(U)$ denotes the number of components and $c(U)$ the number of circuits of U.

Lemma 8.4.4 *Let G be a graph on n vertices with adjacency matrix A. Then*

$$\det A = (-1)^n \sum (-1)^{p(u)} 2^{c(U)},$$

where the summation extends over all spanning basic figures of G.

Proof. The Coates digraph $D(A^*)$ of A is obtained from G by replacing each edge of G with a cycle of two vertices. Because all nonzero entries of A equal 1, the weight of any linear subdigraph of $D(A^*)$ is equal to 1. Each linear subdigraph of $D(A^*)$ can be obtained from a spanning basic figure of G by replacing each isolated edge (i.e., component equal to K_2) by a cycle of two vertices, and by introducing an orientation to each edge of each circuit in one of the two possible ways so that we get a cycle. Therefore, starting from a spanning basic figure U, we can construct $2^{c(U)}$ linear subdigraphs. Now formula (4.1), which defines the determinant of a matrix, reduces to the formula in the lemma. □

We now obtain a formula for the characteristic polynomial of a graph.

Theorem 8.4.5 *The characteristic polynomial of a graph G on n vertices is given by*

$$\sum_{i=0}^{n} a_i \lambda^{n-i},$$

where

$$a_i = \sum_{U_i} (-1)^{p(U_i)} 2^{c(U_i)}, \quad (i = 0, 1, \ldots, n)$$

and the summation extends over all basic figures of G with i vertices.

Proof. If we apply formula (7.2) to the adjacency matrix A of G, we get that a_i is equal to $(-1)^i$ times the sum of the determinants of all the principal submatrices of A of order i. The result now follows from Lemma 8.4.4. □

In the case of a forest, in particular a tree, we obtain a simpler expression for the characteristic polynomial.

Corollary 8.4.6 *The characteristic polynomial of a forest W with n vertices is equal to*

$$\sum_{k=1}^{\left[\frac{n}{2}\right]}(-1)^k m(W,k)\lambda^{n-2k}, \tag{8.5}$$

where $m(W,k)$ is the number of k-matchings in W.

Proof. In a forest, basic figures with an odd number of vertices do not exist. For even $i = 2k$, a basic figure with i vertices is just a k-matching. The corollary now follows from Theorem 8.4.5. □

8.5 Exercises

1. Explain how the Frobenius normal form of a permutation matrix of order n is determined.

2. Determine the Frobenius normal form of the matrix

$$\begin{bmatrix} 0 & 1 & 0 & 0 & 0 & 1 & 0 \\ 0 & 0 & 0 & 0 & 1 & 0 & 0 \\ 0 & 0 & 3 & 0 & 0 & 0 & 1 \\ 0 & 1 & 0 & 0 & 2 & 0 & 0 \\ 0 & 0 & 0 & 1 & 0 & 0 & 2 \\ 1 & 0 & 0 & 0 & 0 & 0 & 0 \\ 0 & 0 & 2 & 0 & 0 & 0 & 0 \end{bmatrix}.$$

3. Determine the Frobenius normal form of the matrix

$$\begin{bmatrix} 0 & 0 & 0 & 1 & 0 & 1 \\ 1 & 2 & 2 & 0 & 0 & 1 \\ 1 & 0 & 1 & 0 & 1 & 1 \\ 2 & 0 & 0 & 0 & 0 & 3 \\ 0 & 3 & 1 & 1 & 0 & 0 \\ 1 & 0 & 0 & 0 & 0 & 3 \end{bmatrix}.$$

4. Show that the following matrices are primitive and determine their exponents:

$$\begin{bmatrix} 0 & 1 & 0 & 0 & 0 & 0 \\ 0 & 0 & 1 & 0 & 0 & 0 \\ 0 & 0 & 0 & 1 & 0 & 0 \\ 0 & 0 & 0 & 0 & 1 & 0 \\ 1 & 0 & 0 & 0 & 0 & 1 \\ 1 & 0 & 0 & 0 & 0 & 0 \end{bmatrix} \text{ and } \begin{bmatrix} 0 & 1 & 0 & 0 & 0 & 0 \\ 0 & 0 & 1 & 0 & 0 & 0 \\ 0 & 0 & 0 & 1 & 0 & 0 \\ 0 & 0 & 0 & 0 & 1 & 0 \\ 1 & 0 & 0 & 0 & 0 & 1 \\ 1 & 1 & 0 & 0 & 0 & 0 \end{bmatrix}.$$

5. Show that a primitive matrix of order $n \geq 2$ contains at least $n+1$ positive entries.

6. Show that if A is primitive, so is A^k for every positive integer k.

7. Let

$$A = \begin{bmatrix} 0 & 1 & 1 & 0 & 0 \\ 0 & 0 & 1 & 0 & 0 \\ 0 & 0 & 0 & 1 & 0 \\ 0 & 0 & 0 & 0 & 1 \\ 1 & 0 & 0 & 0 & 0 \end{bmatrix}.$$

Construct the digraph $D(A)$ and show that A is primitive with exponent 17.

8. Let A be an irreducible nonnegative matrix with at least one positive diagonal element. Prove that A is primitive with exponent at most $2(n-1)$.

9. Use the examples below to show that a nonnegative reducible matrix may or may not have a positive eigenvector:
$$\begin{bmatrix} 1 & 1 \\ 0 & 0 \end{bmatrix}, \quad \begin{bmatrix} 1 & 0 \\ 1 & 0 \end{bmatrix}.$$

10. Determine the Perron root and Perron eigenvector of the matrix
$$\begin{bmatrix} 2 & 3 \\ 1 & 2 \end{bmatrix}.$$

11. Determine the eigenvalues of the graph obtained from the complete graph K_5 by removing an edge.

12. Prove that the largest eigenvalue of a regular graph of degree r is equal to r.

13. Let G be a regular graph of degree r with characteristic polynomial $p(\lambda)$. Determine the characteristic polynomial of the complement $\overline{G}$ of G obtained by joining two vertices by an edge in $\overline{G}$ if and only if they are not joined in G.

14. Check whether the graphs $K_{1,4}$ and $C_4 \cup K_1$ have the same spectrum.[1]

[1]Nonisomorphic graphs that have the same spectrum are called *cospectral* graphs.

Chapter 9

Additional Topics

In this chapter we first introduce some special matrix products (the tensor product and Hadamard product) and prove some of their properties. In Section 9.2, given a square matrix, we show how to determine regions in the complex plane that are sure to contain all of its eigenvalues. In Section 9.3, we introduce an important combinatorial counting function, called the permanent, which, although similar to the determinant in definition, is notoriously difficult to compute in general.

9.1 Tensor and Hadamard Product

Let $A = [a_{ij}]$ and $B = [b_{ij}]$ be matrices of sizes m by p and q by n, respectively. If the number p of columns of A equals the number q of rows of B, then, as we know, A and B can be multiplied to give an m by n matrix whose (i, j)-entry equals

$$\sum_{k=1}^{p} a_{ik}b_{kj} \quad (1 \leq i \leq m; 1 \leq j \leq n).$$

There are other special products of matrices that are often useful in applications.

Definition 9.1.1 The *tensor product* (also called the *Kronecker product*) of A and B in this order is the mq by pn matrix $A \otimes B$

obtained from A by replacing each entry a_{ij} of A with the q by n matrix

$$a_{ij}B \quad (1 \leq i \leq m; 1 \leq j \leq p).$$

The tensor product has a natural partitioned form. If A and B have the same size, that is, $m = q$ and $n = p$, then the *Hadamard–Schur product* or *entrywise product* is the m by n matrix

$$A \circ B = [a_{ij}b_{ij}]$$

obtained by multipying corresponding entries of A and B.

Example 9.1.2 Let

$$A = \begin{bmatrix} 2 & 6 & 3 \\ 4 & 7 & 5 \end{bmatrix} \text{ and } B = \begin{bmatrix} 1 & 4 \\ 5 & 3 \end{bmatrix}.$$

Then

$$\begin{aligned} A \otimes B &= \begin{bmatrix} 2B & 6B & 3B \\ 4B & 7B & 5B \end{bmatrix} \\ &= \left[\begin{array}{cc|cc|cc} 2 & 8 & 6 & 24 & 3 & 12 \\ 10 & 6 & 30 & 18 & 15 & 9 \\ \hline 4 & 16 & 7 & 28 & 5 & 20 \\ 20 & 12 & 35 & 21 & 25 & 15 \end{array}\right]. \end{aligned}$$

If

$$C = \begin{bmatrix} 4 & -3 & 5 \\ -2 & 1 & 6 \end{bmatrix},$$

then

$$A \circ B = \begin{bmatrix} 2(4) & 6(-3) & 3(5) \\ 4(-2) & 7(1) & 5(6) \end{bmatrix} = \begin{bmatrix} 8 & -18 & 15 \\ -8 & 7 & 30 \end{bmatrix}.$$

Assume now that $A = [a_{ij}]$ and $B = [b_{kl}]$ are square matrices of orders m and n, respectively. Let the vertices of the digraph $D(A)$ of A be $\{1, 2, \ldots, m\}$, and let the vertices of the digraph $D(B)$ be $\{1, 2, \ldots, n\}$. Then the digraph $D(A \otimes B)$ has vertices

$$\{(i, j) : 1 \leq i \leq m, 1 \leq j \leq n\}, \tag{9.1}$$

with an edge from a vertex (i, j) to a vertex (k, l) if and only if there is an edge from i to k in $D(A)$ and an edge from j to l in $D(B)$. It is natural to label the rows and columns of $A \otimes B$ with the ordered pairs given in (9.1). With these labels, the rows and columns of $A \otimes B$ occur in the order

$$(1,1), \ldots, (1,n), (2,1), \ldots, (2,n), \ldots, (m,1), \ldots, (m,n).$$

Moreover, the entry $c_{(i,j),(k,l)}$ in position $((i,j),(k,l))$ is given by

$$c_{(i,j),(k,l)} = a_{ik}b_{jl}.$$

The digraph $D(A \otimes B)$ is the *tensor product* (or *Kronecker product*) of the digraphs $D(A)$ and $D(B)$. As a weighted digraph, the edge from (i, j) to (k, l) has weight $a_{ij}b_{kl}$, the product of the weights of the edge from i to k in $D(A)$ and the edge from j to l in $D(B)$.

Although $A \otimes B$ and $B \otimes A$ have the same size, it is not true in general that $A \otimes B = B \otimes A$. For example, if A is the identity matrix and J_2 is the matrix of order 2 each of whose entries equals 1, then

$$I_2 \otimes J_2 = \left[\begin{array}{cc|cc} 1 & 1 & 0 & 0 \\ 1 & 1 & 0 & 0 \\ \hline 0 & 0 & 1 & 1 \\ 0 & 0 & 1 & 1 \end{array}\right] \neq J_2 \otimes I_2 = \left[\begin{array}{cc|cc} 1 & 0 & 1 & 0 \\ 0 & 1 & 0 & 1 \\ \hline 1 & 0 & 1 & 0 \\ 0 & 1 & 0 & 1 \end{array}\right].$$

Although $A \otimes B \neq B \otimes A$ in general, the following theorem about their digraphs does hold.

Theorem 9.1.3 *There exists a permutation matrix P of order mn such that*

$$P(A \otimes B)P^T = B \otimes A.$$

Equivalently, the digraphs $D(A \otimes B)$ and $D(B \otimes A)$ are isomorphic with the isomorphism preserving weight.

Proof. In the weighted digraph $D(A \otimes B)$ there is an edge from (i, j) to (k, l) of weight $a_{ik}b_{jl}$. Similarly, in the weighted digraph $D(B \otimes A)$ there is an edge from (j,i) to (l,k) of weight $b_{jl}a_{ik} = a_{ik}b_{jl}$. Thus the bijection

$$((i,k),(j,l)) \to ((j,l),(i,k))(1 \leq i,j \leq m, 1 \leq k,l \leq n)$$

is an isomorphism of the two weighted digraphs. □

In the following theorem we collect a number of elementary identities for the tensor product.

Theorem 9.1.4 *If A, B, and C are matrices of appropriate sizes in order to carry out the indicated operations, the following hold:*

(i) *(associative rule)* $A \otimes (B \otimes C) = (A \otimes B) \otimes C$.

(ii) *(distributive rule)* $A \otimes (B + C) = A \otimes B + A \otimes C$.

(iii) *(distributive rule)* $(A + B) \otimes C = A \otimes C + B \otimes C$.

(iv) *(transpose rule)* $(A \otimes B)^T = A^T \otimes B^T$.

(v) *(product rule)* $(A \otimes B)(C \otimes D) = AC \otimes BD$.

Proof. Identities (i)–(iv) can be verified in a straightforward manner. We now verify (v). In order for (v) to make sense, the number of columns of A has to equal the number of rows of C, and the number of columns of B has to equal the number of rows of D.

The entry in position $((i,j),(k,l))$ of $(A \otimes B)(C \otimes D)$ equals

$$\sum_p \sum_q a_{ip}b_{jq} \cdot c_{pk}d_{ql} = \sum_p a_{ip}c_{pk} \cdot \sum_q b_{jq}d_{ql},$$

and this is the same as the entry in the $((i,j),(k,l))$ position of $AC \otimes BD$. □

Note that if A and B are square matrices of orders m and n, respectively, then (v) implies that

$$A \otimes B = (A \otimes I_n)(I_m \otimes B).$$

The product rule (v) in Theorem 9.1.4 has some useful and, in some cases, surprising consequences for the tensor product of square matrices.

Theorem 9.1.5 *Let A and B be square matrices of orders m and n, respectively.*

Then the following hold:

(i) *If A and B are invertible, then $A \otimes B$ is invertible and $(A \otimes B)^{-1} = A^{-1} \otimes B^{-1}$.*

(ii) *If $\lambda_1, \lambda_2, \ldots, \lambda_m$ are the eigenvalues of A and $\mu_1, \mu_2, \ldots, \mu_n$ are the eigenvalues of B, then the eigenvalues of $A \otimes B$ are the mn products of the eigenvalues of A with the eigenvalues of B:*

$$\lambda_i \mu_j \quad (1 \leq i \leq m, 1 \leq j \leq n).$$

(iii) $\det(A \otimes B) = (\det A)^n (\det B)^m$.

Proof. To establish (i) we use the product rule for tensor products to compute

$$(A \otimes B)(A^{-1} \otimes B^{-1}) = (AA^{-1}) \otimes (BB^{-1}) = I_m \otimes I_n = I_{mn}.$$

Thus $A^{-1} \otimes B^{-1}$ is the inverse of $A \otimes B$. It is easy to establish that if λ is an eigenvalue of A and μ is an eigenvalue of B, then $\lambda\mu$ is an eigenvalue of $A \otimes B$. We simply choose an eigenvector $u \neq 0$ of A for λ and an eigenvector $v \neq 0$ of B for μ. Then $u \otimes v$ is not the zero vector and, by the product rule,

$$(A \otimes B)(u \otimes v) = (Au) \otimes (Bv) = (\lambda u) \otimes (\mu v) = \lambda\mu(u \otimes v).$$

In order to know that the entire collection of eigenvalues of $A \otimes B$ is as given in (ii) (that is, that the multiplicities work out), we use the Jordan canonical forms J_A and J_B of A and B, respectively. There exist invertible matrices P and Q such that $P^{-1}AP = J_A$ and $QBQ^{-1} = J_B$, and, by (i), $P \otimes Q$ is invertible with $(P \otimes Q)^{-1} = P^{-1} \otimes Q^{-1}$. Using the product rule, we get that

$$\begin{aligned}(P \otimes Q)^{-1}(A \otimes B)(P \otimes Q) &= (P^{-1} \otimes Q^{-1})(A \otimes B)(P \otimes Q) \\ &= (P^{-1}AP) \otimes (Q^{-1}BQ) \\ &= J_A \otimes J_B.\end{aligned}$$

Now J_A is a triangular matrix with $\lambda_1, \lambda_2, \ldots, \lambda_m$ on its main diagonal, and J_B is a triangular matrix with $\mu_1, \mu_2, \ldots, \mu_n$ on its main diagonal. The matrix $J_A \otimes J_B$ is then a triangular matrix with the mn numbers $\lambda_i\mu_j$ $(1 \leq m \leq n, 1 \leq j \leq n)$ on its main diagonal. Because the eigenvalues of a triangular matrix are its diagonal entries, this establishes (ii).

To prove (iii), we simply note that $\det A = \lambda_1\lambda_2\cdots\lambda_m$ and $\det B = \mu_1\mu_2\cdots\mu_n$, and use (ii) to conclude that

$$\begin{aligned}\det A \otimes B &= \prod_{i=1}^{m}\prod_{j=1}^{n} \lambda_i\mu_j \\ &= (\lambda_1\lambda_2\cdots\lambda_m)^n(\mu_1\mu_2\cdots\mu_n)^m \\ &= (\det A)^n(\det B)^m.\end{aligned}$$

□

We only briefy discuss the Hadamard–Schur product $A \circ B = [a_{ij}b_{ij}]$ of two square matrices $A = [a_{ij}]$ and $B = [b_{ij}]$ of order n.

The matrix $A \circ B$ is a principal submatrix of the tensor product $A \otimes B$. In fact, using our labeling of the rows and columns of $A \otimes B$, if we let $K = \{(1,1), (2,2), \ldots, (n,n)\}$, then $A \circ B$ is the principal submatrix $(A \otimes B)[K, K]$ of $A \otimes B$. The weighted digraph $D(A \circ B)$ is obtained from the weighted digraphs of A and B by multiplying the corresponding weights. In particular, there is an edge from vertex i to vertex j of nonzero weight, if and only if there is an edge from i to j of nonzero weight in both $D(A)$ and $D(B)$. In unweighted terms, the edges of $D(A \circ B)$ are the edges common to $D(A)$ and $D(B)$.

9.2 Eigenvalue Inclusion Regions

Usually the eigenvalues of a square matrix cannot be determined exactly and, as a result, it is useful to determine regions in the complex plane that include all the eigenvalues and which can easily be computed. The first theorem giving an eigenvalue inclusion region is the theorem of Geršgorin proved in 1931. To state this theorem we require a few preliminaries.

Let $A = [a_{ij}]$ be a matrix of order n. Let

$$r_i(A) = \sum_{j \neq i} |a_{ij}| \quad (i = 1, 2, \ldots, n)$$

be the sum of the absolute values of the entries in row i with the entry a_{ii} on the main diagonal deleted. In addition, let

$$\Gamma_i(A) = \{z : z \text{ a complex number}, |z - a_{ii}| \leq r_i(A)\}$$

be the disk in the complex plane centered at a_{ii} with radius $r_i(A)$, called the ith *Geršgorin disk* of A. Finally, let

$$\Gamma(A) = \cup_{i=1}^{n} \Gamma_i(A)$$

be the union of all the Geršgorin disks of A. Then $\Gamma(A)$ is a union of disks in the complex plane and is called the *Geršgorin region* of A.

Example 9.2.1 Let

$$A = \begin{bmatrix} 2 & 4 \\ 3 & i \end{bmatrix}.$$

Then $\Gamma_1(A)$ is the disk centered at the point $(2, 0)$ on the real axis with radius 4 and $\Gamma_2(A)$ is the disk centered at the point $(0, 1)$ on the imaginary axis with radius 3. Now let

$$A = \begin{bmatrix} 1 & 3 & 5 \\ 2 & 0 & 4 \\ -1 & 3 & -2 \end{bmatrix}.$$

Then $\Gamma_1(A)$, $\Gamma_2(A)$, $\Gamma_3(A)$ are, respectively, the disks centered at $(1, 0)$, $(0, 0)$, and $(-2, 0)$ with radii 8, 6, and 4, respectively. □

The theorem of Geršgorin is that the Geršgorin region of a matrix contains all its eigenvalues.

Theorem 9.2.2 *Let $A = [a_{ij}]$ be a matrix of order n. Then all the eigenvalues of A are contained in its Geršgorin region $\Gamma(A) = \cup_{i=1}^{n} \Gamma_i(A)$.*

Proof. Let λ be an eigenvalue of A and let $x = [x_1\ x_2\ \ldots x_n]^T$ be an eigenvector of A corresponding to λ. Then $Ax = \lambda x$ implies that

$$\sum_{j=1}^{n} a_{ij} x_i = \lambda x_i \quad (i = 1, 2, \ldots, n). \tag{9.2}$$

Because x is an eigenvector, x has an entry different from zero. We choose k so that

$$|x_k| = \max\{|x_1|, |x_2|, \ldots, |x_n|\}.$$

Then $|x_k| > 0$, and we consider the kth equation in (9.2) and obtain

$$\sum_{j=1}^{n} a_{kj} x_j = \lambda x_k.$$

Rewriting this equation by grouping together the two coefficients of x_k, we get

$$(\lambda - a_{kk}) x_k = \sum_{j \neq k} a_{kj} x_j.$$

We now take the absolute value of both sides, and, using the triangle inequality, we obtain

$$\begin{aligned} |\lambda - a_{kk}||x_k| &= |\sum_{j \neq k} a_{kj} x_j| \\ &\leq \sum_{j \neq k} |a_{kj}||x_j| \\ &\leq \sum_{j \neq k} |a_{kj}||x_k| \\ &= r_k(A)|x_k|. \end{aligned}$$

Because $|x_k| > 0$, we obtain upon cancellation that

$$|\lambda - a_{kk}| \leq r_k(A).$$

Thus λ is in the kth Geršgorin disk, and hence in the Geršgorin region. □

As a corollary, we obtain an upper bound on the largest absolute value of an eigenvalue of a matrix.

Corollary 9.2.3 *Let $A = [a_{ij}]$ be a matrix of order n. Let*

$$r(A) = \max\{\sum_{j=1}^{n} |a_{ij}| : 1 \leq i \leq n\},$$

the maximum of the sum of the absolute values of the entries in each row of A. Then all the eigenvalues of A are contained in the disk

$$\{z : z \text{ a complex number}, |z| \leq r(A)\}.$$

Proof. Let λ be an eigenvalue of A. By Theorem 9.2.2, there exists a k such that λ is in the kth Geršgorin disk:

$$|\lambda - a_{kk}| \leq r_k(A) = \sum_{i \neq k} |a_{ki}|.$$

Because for complex numbers x and y, $|x| - |y| \leq |x - y|$, we obtain

$$|\lambda| - |a_{kk}| \leq \sum_{i \neq k} |a_{ki}|,$$

that is,

$$|\lambda| \leq \sum_{i=1}^{n} |a_{ki} \leq r(A).$$

□

Theorem 9.2.2 implies a result that gives a sufficient condition for the invertibility of a matrix. We state this as another corollary.

Call a matrix $A = [a_{ij}]$ *diagonally dominant* provided

$$|a_{ii}| > r_i(A) = \sum_{j \neq i} |a_{ij}|. \qquad (9.3)$$

Corollary 9.2.4 *Let $A = [a_{ij}]$ be a diagonally dominant matrix of order n. Then A is invertible.*

Proof. We know that a matrix is invertible if and only if 0 is not an eigenvalue of A. The diagonal dominance condition (9.3) implies that none of the Geršgorin disks

$$\Gamma_i(A) = \{z : z \text{ a complex number}, |z - a_{ii}| \leq r_i(A)\}$$

contains the complex number 0. Because by Theorem 9.2.2 the eigenvalues of A are contained in the union of these Geršgorin disks, 0 is not an eigenvalue of A and A is invertible. □

It is natural to ask about the boundary of the Geršgorin region $\Gamma(A)$ of A. We state without proof the following theorem of Taussky.

Theorem 9.2.5 *Let A be an irreducible matrix of order n, that is, the digraph $D(A)$ is strongly connected. Then, if an eigenvalue λ lies on the boundary of the the Geršgorin region, then λ is on the boundary of every one of the n Geršgorin disks, that is,*

$$|\lambda - a_{ii}| = r_i(A) \quad (i = 1, 2, \ldots, n).$$

We now turn to showing how knowledge of the digraph of a matrix can be used to obtain refined eigenvalue inclusion regions. We need a few preliminaries.

Let D be a digraph with vertex set $\{1, 2, \ldots, n\}$, where there is a *weight* w_i associated with each vertex i ($1 \leq i \leq n$). Thus D is a *vertex-weighted digraph.* Let u be a vertex of D and let (u, v) be an edge leaving u. Then (u, v) is a *dominant edge* from u provided that for each edge (u, x) leaving u, we have $w_v \geq w_x$. Thus the edge (u, v) is a dominant edge from u provided there is no edge from u to a vertex of larger weight than the weight of v. Now consider a cycle $\gamma = a_1, a_2, \ldots, a_k, a_1$. Then γ is a *dominant cycle* in D provided that each of its edges $(a_1, a_2), \ldots, (a_{k-1}, a_k), (a_k, a_1)$ is a dominant edge.

Lemma 9.2.6 *Let D be a vertex-weighted digraph such that each vertex has a positive outdegree. Then D has a dominant cycle.*

Proof. We start at any vertex x_1 of D and choose a dominant edge (x_1, x_2) leaving x_1. Then we choose a dominant edge (x_2, x_3) leaving x_2. We continue like this until we first repeat a vertex, say vertex x_k, thereby obtaining a cycle

$$\gamma = x_k, x_{k+1}, \ldots, x_p = x_k.$$

Then γ is a dominant cycle. □

Now let $A = [a_{ij}]$ be a matrix of order n and consider the quantities

$$r_i(A) = \sum_{j \neq i} |a_{ij}| \quad (i = 1, 2, \ldots, n).$$

Let $D_0(A)$ be the digraph obtained by removing all loops (cycles of length 1) from the digraph $D(A)$. We regard $D_0(A)$ as a vertex-weighted digraph where the weight of each vertex i equals $r_i(A)$. We may also regard $D_0(A)$ as a vertex-weighted digraph where the weight of vertex i is $|a_{ii}|$ $(i = 1, 2, \ldots, n)$. We then have the following theorem, which is a generalization of Corollary 9.2.4. (We shall reverse the order above in which the eigenvalue inclusion region given by Theorem 9.2.2 gave a condition (diagonal dominance) for a matrix to be invertible by proving first an invertibility theorem and obtaining from it an eigenvalue inclusion region.)

If γ is a cycle of a vertex-weighted digraph, then by $\prod_\gamma w_i$ we mean the product of the weights of all the vertices i of γ.

Theorem 9.2.7 *Let $A = [a_{ij}]$ be a matrix of order n each of whose entries on the main diagonal is different from zero. Assume that*

$$\prod_\gamma |a_{ii}| > \prod_\gamma r_i(A) \tag{9.4}$$

for all cycles γ of $D(A)$ of length at least 2. Then A is an invertible matrix.

Proof. The graphs $D(A)$ and $D_0(A)$ have the same cycles of length at least 2. We show that A is invertible by showing that $\det A \neq 0$. The determinant of a matrix is the product of the determinants of its irreducible components. Because we are assuming that the entries on the main diagonal of A are different from zero, we assume that A is an irreducible matrix of order $n \geq 2$; thus $D_0(A)$ is a strongly connected digraph with at least two vertices.

Assume to the contrary that $\det A = 0$. Then the rank of A is strictly less than n and there is a nonzero vector $x = [x_1\ x_2\ \ldots x_n]$ such that $Ax = 0$. Let $I = \{i : x_i \neq 0\}$ and let $A' = A[I, I]$ be the principal submatrix of A obtained by deleting those rows and columns whose index does not belong to I. Let x' be obtained from x in a similar way. Then each coordinate x'_j of x' is different from zero and

$$A'x' = 0. \tag{9.5}$$

Because each entry on the main diagonal of A' is nonzero, (9.5) implies that each vertex of $D_0(A')$ has an edge leaving it. We weight the vertices of $D_0(A')$, that is, those i in I, by $|x_i|$ and apply Lemma 9.2.6 to obtain a dominant cycle $\gamma = i_1, i_2, \ldots, i_p, i_{p+1} = i_p$ of $D_0(A')$ of length $p \geq 2$. From (9.5) we get that for $1 \leq j \leq p$,

$$a_{i_j i_j} x_{i_j} = - \sum_{k \in I \setminus \{i_j\}} a_{i_j k} x_k.$$

Using the triangle inequality and the fact that γ is a dominant cycle in $D_0(A')$ we obtain

$$\begin{aligned} |a_{i_j i_j}||x_{i_j}| &\leq \sum_{k \in I \setminus \{i_j\}} |a_{i_j k}||x_k| \\ &\leq \left(\sum_{k \in I \setminus \{i_j\}} |a_{i_j k}| \right) |x_{i_{k+1}}| \\ &\leq r_{i_j}(A)|x_{i_{j+1}}|. \end{aligned}$$

Multiplying the last inequalities for $j = 1, 2, \ldots, p$ we obtain

$$\prod_{\gamma} |a_{ii}| \prod_{\gamma} |x_j| \leq \prod_{\gamma} r_j(A) \prod_{\gamma} |x_j|,$$

and because $x_j \neq 0$ for j in I,

$$\prod_{\gamma} |a_{ii}| \leq \prod_{\gamma} r_j(A).$$

This last inequality contradicts (9.4), and the proof is complete. □

We remark that if the matrix A in Theorem 9.2.7 is irreducible, the assumption that the entries on the main diagonal are different from zero is implied by (9.4). This is because $D_(A)$ is then strongly connected and every vertex is on a cycle.

We now obtain the eigenvalue inclusion region corresponding to Theorem 9.2.7. For simplicity, we assume that our matrix is an irreducible matrix of order at least 2. For each cycle γ of $D(A)$ of length at least 2, we define the *lemniscate*

$$Z_\gamma = \left\{ z : \prod_\gamma |z - a_{ii}| \leq \prod_\gamma r_i(A) \right\}.$$

Theorem 9.2.8 *Let $A = [a_{ij}]$ be an irreducible matrix of order $n \geq 2$. Then all the eigenvalues of A are included in the region of the complex plane specified by the union of Z_γ taken over all cycles γ of $D(A)$ of length at least 2.*

Proof. Let λ be an eigenvalue of A and consider the singular matrix $\lambda I_n - A$. The graphs $D(A)$ and $D(\lambda I_n - A)$ have the same cycles of length at least 2, and $r_i(A) = r_i(\lambda I_n - A)$ for each $i = 1, 2, \ldots, n$. Because the matrix $\lambda I_n - A$ is singular, it follows from Theorem 9.2.7 that there is a cycle γ of $D(A)$ of length at least 2 such that

$$\prod_\gamma |\lambda - a_{ii}| \leq \prod_\gamma r_i(A).$$

Thus λ is in Z_γ and the theorem holds. □

Special cases of Theorems 9.2.7 and 9.2.8 are contained in the following theorem of Brauer. We leave it as an exercise to provide the proof.

Corollary 9.2.9 *Let $A = [a_{ij}]$ be a matrix of order $n \geq 2$. If*

$$|a_{ii}a_{jj}| > r_i(A)r_j(A) \quad (1 \leq i < j \leq n),$$

then A is invertible. The eigenvalues of A are all contained in the region of the complex plane specified by the union of the ovals

$$Z_{ij} = \{z : |z - a_{ii}||z - a_{jj}| \leq r_i(A)r_j(A) \quad (1 \leq i < j \leq n).$$

By use of the digraph, we were able to give a substantial generalization of the Geršgorin inclusion region for the eigenvaues of a matrix. One can consult [7] and [78] for a lot more on this topic, including proofs of results not given here.

9.3 Permanent and SNS-Matrices

Let $A = [a_{ij}]$ be a matrix of order n. The definition of the *permanent* of A follows the classical definition of the determinant given in Theorem 4.4.2 but with a simplification. Ironically, this simplification in the formula makes it more difficult to compute the permanent.

The permanent of A is the number given by the formula

$$\operatorname{per} A = \sum_{(j_1,j_2,\ldots,j_n)\in S_n} a_{1j_1}a_{2j_2}\cdots a_{nj_n}, \tag{9.6}$$

where the summation is over all permutations $(j_1, j_2, \ldots, j_n)$ of $\{1, 2, \ldots, n\}$. Thus, unlike the determinant, we don't put a minus sign in front of some of the terms in the summation in (9.6). In the permanent we compute all possible products of n entries of A provided these n entries come from different rows and different columns. As a result, the permanent does not change if we permute the rows of A and permute the columns of A. In addition, the permanent does not change when a matrix is transposed. An equivalent way to define the permanent uses the weighted Coates digraph $D^*(A)$. Recall that, according to Definition 4.1, the determinant of A is given by

$$\det A = (-1)^n \sum_{L\in\mathcal{L}(A)} (-1)^{c(L)} w(L)$$

where the summation is over all linear subdigraphs of $D^*(A)$ and the weight $w(L)$ of L equals the product of the weights of its edges. The corresponding formula for the permanent is the simpler

$$\operatorname{per} A = \sum_{L\in\mathcal{L}(A)} w(L).$$

We record the basic, easily verifiable, properties of the permanent in the next lemma and leave their verification to the reader.

Lemma 9.3.1 *The following properties hold for a matrix A of order n:*

(i) $\operatorname{per} PAQ = \operatorname{per} A$ *for all permutation matrices* P *and* Q *of order* n.

(ii) $\operatorname{per} A^T = \operatorname{per} A$.

(iii) $\operatorname{per} cA = c^n \operatorname{per} A$ *for all scalars* c.

(iv) *If* A *has a row (or column) of all zeros, then* $\operatorname{per} A = 0$.

(v) *If some row (or some column) of* A *is multiplied by a scalar* c*, then the permanent of the resulting matrix equals* $c\operatorname{per} A$.

(vi) $\operatorname{per} P = 1$ *for every permutation matrix* P *of order* n. *In particular,* $\operatorname{per} I_n = 1$.

(vii) *If* $A = B \oplus C$*, where* B *and* C *are square matrices, then* $\operatorname{per} A = \operatorname{per} B \operatorname{per} C$. *More generally, if*

$$A = \begin{bmatrix} B & O \\ X & C \end{bmatrix},$$

where A *and* B *are square matrices, then* $\operatorname{per} A = \operatorname{per} B \mathit{per} C$.

(viii) *(Laplace expansion by a row or column)*

$$\operatorname{per} A = \sum_{j=1}^{n} a_{ij} \operatorname{per} A_{i,j} \quad (j = 1, 2, \ldots, n)$$

and

$$\operatorname{per} A = \sum_{i=1}^{n} a_{ij} \operatorname{per} A_{i,j} \quad (i = 1, 2, \ldots, n).$$

(Recall that $A_{i,j}$ *is the matrix of order* $n-1$ *obtained from* A *by striking out row* i *and column* j*.)* □

Example 9.3.2 Let

$$A = \begin{bmatrix} a & b \\ c & d \end{bmatrix}.$$

Then $\operatorname{per} A = ad + bc$. Let

$$B = \begin{bmatrix} a & b & 0 \\ 0 & c & d \\ e & 0 & f \end{bmatrix}.$$

Then it is easy to see that in the permanent of B there are at most two nonzero terms, namely, acf and bde. Hence $\operatorname{per} B = acf + bdf$. Now let

$$C = \begin{bmatrix} 1 & 2 & 2 \\ 3 & 1 & 1 \\ 1 & 2 & 1 \end{bmatrix}.$$

Then each of the $3! = 6$ terms in the permanent of C is nonzero, and adding them we obtain that

$$\operatorname{per} C = 1 + 2 + 12 + 2 + 6 + 2 = 25.$$

Let D be the matrix obtained from C by adding the second row to the first row. Then

$$D = \begin{bmatrix} 4 & 3 & 3 \\ 3 & 1 & 1 \\ 1 & 2 & 1 \end{bmatrix}.$$

A simple calculation shows that $\operatorname{per} D = 45$. This example shows that the elementary row operation of adding a multiple of one row to another row can change the value of the permanent. In the case of the determinant, we have $\det D = \det C$. The fact that such elementary row (or column) operations can alter the permanent leads to the general computational difficulty in evaluating the permanent. $\square$

Let $A = [a_{ij}]$ be a square matrix of order n. Recall that the König digraph $G(A)$ of A has n black vertices corresponding to the rows of A and n white vertices corresponding to the columns of A, and an edge from black vertex i to white vertex vertex j of weight a_{ij}. Recall (see Section 4.4) also that a collection F of n edges, one leaving each black vertex and simultaneously one entering each white vertex, is a 1-factor (or perfect matching) of $G(A)$ and its weight $w(F)$ is the product of weights of these n edges. If $\mathcal{F}(A)$ denotes the collection of all 1-factors of $G(A)$, then it follows that

$$\operatorname{per} A = \sum_{F \in \mathcal{F}(A)} w(F),$$

the sum of the weights of all the 1-factors of $G(A)$.

As usual, when $a_{ij} = 0$, we can consider that there is no edge from black vertex i to white vertex j, that is, weight equal zero is interpreted as the absence of an edge, and thus edges of weight zero are not part of any 1-factor. Now consider the special case where each entry of A equals 0 or 1, that is, A is a $(0,1)$-matrix. Then $w(F) = 0$ or 1 and so, with our convention, the permanent of A counts the number of 1-factors of A. Thus the permanent is a counting function and, indeed, one of some significance.

Another way to view the permanent of a (0,1)-matrix A is as the number of permutation matrices P of order n such that $P \leq A$ (entrywise). This is so because each such permutation matrix $P \leq A$ corresponds to a 1-factor of weight 1 and viceversa.

Example 9.3.3 Let A be the matrix of order n having 0's everywhere on its main diagonal and 1's everywhere off the main diagonal. Thus $A = J_n - I_n$, where J_n is the matrix of order n of all 1's. The permutation matrices P with $P \leq A$ (entrywise) correspond to those permutations $i_1 i_2 \ldots i_n$ of $\{1, 2, \ldots, n\}$ such that $i_k \neq k$ for $k = 1, 2, \ldots, n$. Such permutations are called *derangements* (of order n) since in such a permutation matrix, no integer is in its natural position (the natural position for integer k is position k, and this is precluded under the assumption that $i_k \neq k$). The number of derangements is denoted by D_n and thus we have $\operatorname{per}(J_n - I_n) = D_n$. We easily calculate that $D_1 = 0$, $D_2 = 1$, and $D_3 = 2$ (the permutations $2, 1, 3$ and $3, 1, 2$ are the derangements of order 3). The inclusion-exclusion formula (see Section 1.3) can be used to count the number of permutations of order n as follows. Let X_k be the set of permutations $i_1 i_2 \ldots i_n$ of $\{1, 2, \ldots, n\}$ in which $i_k = k$ $(k = 1, 2, \ldots, n)$. Then the derangements are those permutation in the intersection $\overline{X_1} \cap \overline{X_2} \cap \cdots \cap \overline{X_n}$ of the complements of the X_k's in the set of all permutations of

$\{1, 2, \ldots, n\}$. Thus

$$\begin{aligned}
\operatorname{per}(J_n - I_n) &= |\overline{X_1} \cap \overline{X_2} \cap \cdots \cap \overline{X_n}| \\
&= \sum_{k=0}^{n} \sum_{K \subset \{1,2,\ldots,n\}; |K|=k} (-1)^{|K|} |\cap_{i \in K} X_i| \\
&= \sum_{k=0}^{n} (-1)^k \binom{n}{k} |X_1 \cap X_2 \cap \cdots \cap X_k| \\
&= \sum_{k=0}^{n} (-1)^k \binom{n}{k} (n-k)! \\
&= \sum_{k=0}^{n} (-1)^k \frac{n!}{k!} \\
&= n! \sum_{k=0}^{n} (-1)^k \frac{1}{k!}.
\end{aligned}$$

Here we have used the fact that $|\cap_{i \in K} X_i|$ depends only on the cardinality k of K and thus equals $|X_1 \cap X_2 \cap \cdots \cap X_k|$. Since $|X_1 \cap X_2 \cap \cdots \cap X_k|$ counts the number of permutations of the form $12 \ldots k i_{k+1} \ldots i_n$, its value is $(n-k)!$. As an example of this formula, we calculate that

$$D_4 = \operatorname{per}(J_4 - I_4) = 4! \left(1 - \frac{1}{1!} + \frac{1}{2!} - \frac{1}{3!} + \frac{1}{4!}\right) = 9.$$

The resemblance of the permanent to the determinant naturally leads one to the question of whether it might be possible to use the determinant in order to calculate the permanent.

Example 9.3.4 Let

$$A = \begin{bmatrix} a & b \\ c & d \end{bmatrix},$$

and let A' be the matrix obtained from A by attaching a minus sign to the entry b:

$$A' = \begin{bmatrix} a & -b \\ c & d \end{bmatrix}.$$

Then $\det A' = ac - (-b)d = ac + bd = \operatorname{per} A$. Thus, attaching the minus sign to b *converts* the determinant into the permanent; the determinant of the resulting matrix A' equals the permanent of the original matrix A, no matter what the values of $a, b, c,$ and d. In other words, the identity $\det A' = \operatorname{per} A$ is an *algebraic identity.* Now let

$$A = \begin{bmatrix} a & b & c \\ d & e & f \\ g & h & i \end{bmatrix}.$$

Can we attach minus signs to some of the entries of A in order to convert the determinant into the permanent? The matrix of order 3 has the property that, in the classical formula for the determinant, both the even permutations and the odd permutations partition the entries of A:

$$\text{(even permutation terms)} \quad aei, \ bfg, \ \text{and } cdh; \tag{9.7}$$

$$\text{(odd permutation terms)} \quad ceg, \ bdi, \ \text{and } ahf. \tag{9.8}$$

If we are to convert the determinant into the permanent, then the terms in (9.7) must each have an even number of minus signs in them while the terms in (9.8) must each have an odd number of signs in them. Since the sum of three odd numbers is odd and the sum of three even numbers is even, this is impossible. Thus the determinant of the general matrix of order 3 cannot be converted into its permanent.

Now suppose we assume that c is identically zero. Thus A now takes the form

$$A = \begin{bmatrix} a & b & 0 \\ d & e & f \\ g & h & i \end{bmatrix},$$

and the permanent of A satisfies

$$\operatorname{per} A = aei + bfg + bdi + ahf.$$

Let

$$A' = \begin{bmatrix} a & -b & 0 \\ d & e & -f \\ g & h & i \end{bmatrix}.$$

Then calculating the determinant of A' gives the algebraic identity

$$\det A' = aei + bfg + bdi + ahf = \operatorname{per} A. \tag{9.9}$$

Thus, affixing minus signs to b and f converts the determinant into the permanent. Because (9.9) is an algebraic identity, we can reformulate our discussion as follows. Let

$$B = \begin{bmatrix} 1 & 1 & 0 \\ 1 & 1 & 1 \\ 1 & 1 & 1 \end{bmatrix}.$$

There are four nonzero terms in the determinant and permanent of B. All these terms in the permanent have value 1. In the determinant, two have value 1 (corresponding to the even permutations $1, 2, 3$ and $2, 3, 1$) and two have value -1 (corresponding to the odd permutations $2, 1, 3$ and $1, 3, 2$). To convert the determinant into the permanent we need to change some of the 1's to -1's in order that all the nonzero terms in the determinant now have value 1 (*and so no cancellation occurs*). This is accomplished with the matrix

$$B' = \begin{bmatrix} 1 & -1 & 0 \\ 1 & 1 & -1 \\ 1 & 1 & 1 \end{bmatrix},$$

where

$$\det B' = 4 = \operatorname{per} B.$$

The examples above lead to the following definition.

Definition 9.3.5 Let A' be a $(0, 1, -1)$-matrix, that is, a matrix each of whose entries is 0, 1, or -1 with at least one nonzero term in its classical determinant expansion. Let A be the matrix obtained from A' by replacing each of its -1's with 1's. Then A' is a *sign-nonsingular matrix* (abbreviated SNS-matrix) provided that $\det A' = \pm \operatorname{per} A$. If A' is an SNS-matrix, then, in evaluating the determinant of A' using the classical expansion, there can be no cancellation of nonzero terms; either all the nonzero terms equal 1 (so $\det A = \operatorname{per} A$) or all the nonzero terms have value -1 (so $\det A = -\operatorname{per} A$).

Proceeding in the other direction we get the following. Start with a $(0, 1)$-matrix $A = [a_{ij}]$ of order n such that the permanent of A is not zero (the König digraph has a perfect matching). If an

SNS-matrix A' can be obtained from A by changing some of the 1's of A to -1's, then $\det A' = \pm \operatorname{per} A$, and we have succeeded in *converting the determinant into the permanent*, or we might better say, that we have succeeded in *converting the permanent of A into a determinant.* We note that in case we get $\det A' = -\operatorname{per} A$, then by multiplying the entries of A' in row 1 by -1, we obtain an SNS-matrix A'' with $\det A'' = \operatorname{per} A$. As a result we can ignore, with no loss in generality, the possibility that $\det A' = -\operatorname{per} A$.

Example 9.3.6 The following matrices are SNS-matrices:

$$\begin{bmatrix} 1 & -1 \\ 1 & 1 \end{bmatrix}, \begin{bmatrix} 1 & -1 & 0 \\ 1 & 1 & -1 \\ 1 & 1 & 1 \end{bmatrix}, \begin{bmatrix} 1 & -1 & 0 & 0 \\ 1 & 1 & -1 & 0 \\ 1 & 1 & 1 & -1 \\ 1 & 1 & 1 & 1 \end{bmatrix}.$$

The example of order 4 shows that the permanent of the matrix

$$\begin{bmatrix} 1 & 1 & 0 & 0 \\ 1 & 1 & 1 & 0 \\ 1 & 1 & 1 & 1 \\ 1 & 1 & 1 & 1 \end{bmatrix}$$

can be converted into a determinant; the permanent equals 8. Moreover, because we get no cancellation of nonzero terms in the determinant, we obtain the algebraic identity that for

$$B = \begin{bmatrix} a & b & 0 & 0 \\ c & d & e & 0 \\ f & g & h & i \\ j & k & l & m \end{bmatrix}$$

and

$$B' = \begin{bmatrix} a & -b & 0 & 0 \\ c & d & -e & 0 \\ f & g & h & -i \\ j & k & l & m \end{bmatrix},$$

we have

$$\det B' = \operatorname{per} B.$$

We collect some elementary properties of SNS-matrices in the following lemma.

Lemma 9.3.7 *Let A be an SNS-matrix of order n. Then the following hold:*

(i) *A has a nonzero term in its classical determinant expansion.*

(ii) *A^T is an SNS-matrix.*

(iii) *If P and Q are permutation matrices of order n, then PAQ is an SNS-matrix.*

(iv) *Every matrix obtained from A by multiplying some rows and columns by -1's is an SNS-matrix. Equivalently, if D_1 and D_2 are diagonal matrices with only 1's and -1's on the main diagonal, then D_1AD_2 is an SNS-matrix.* □

We now discuss an important connection between SNS-matrices and digraphs. Let $A = [a_{ij}]$ be a $(0, 1, -1)$-matrix of order n. We consider under what circumstances A is an SNS-matrix. In order for A to have a chance of being SNS, it must have a nonzero term in its determinant expansion (see (i) of Lemma 9.3.7). By (iii) of Lemma 9.3.7, we may assume that this nonzero term is the product of the entries on the main diagonal, that is, all entries on the main diagonal of A are nonzero. By (iv) of Lemma 9.3.7, we may further assume that all entries on the main diagonal equal -1. With these assumptions, we now consider the weighted digraph $D(A)$. Each edge of $D(A)$ has weight ± 1 (as we often do, we ignore edges of weight 0).

Theorem 9.3.8 *Let $A = [a_{ij}]$ be a $(0, 1, -1)$-matrix of order n each of whose entries on the main diagonal equals -1. Then A is an SNS-matrix if and only if the weight of each cycle in the Coates digraph $D^*(A)$ equals -1.*

Proof. We refer to the definition of the determinant given in (4.1):

$$\det(A) = (-1)^n \sum_{L \in \mathcal{L}(A)} (-1)^{c(L)} w(L), \tag{9.10}$$

where the summation extends over all linear subdigraphs L of the Coates digraph $D^*(A)$; we may restrict this summation to those Ls for which $w(L) \neq 0$. The term corresponding to the linear subdigraph consisting of n loops, one at each vertex, of weight -1 (corresponding to the identity permutation $1, 2, \ldots, n$) equals

$$(-1)^n a_{11}a_{22}\cdots a_{nn} = (-1)^n(-1)^n = (-1)^{2n} = 1.$$

The matrix A is an SNS-matrix if and only if all nonzero terms $(-1)^{c(L)}w(L)$ in the summation (9.10) equal 1.

First suppose that A is an SNS-matrix. The weights of the cycles of length 1, the loops, equal -1 because A has all -1's on its main diagonal. Let γ be a cycle of length $k \geq 2$, and let L be the linear subdigraph whose cycles are γ and the $n-k$ loops at the vertices are not contained on γ. Then $c(L) = 1 + n - k$ and $w(L) = w(\gamma) \cdot (-1)^{n-k}$. Because A is an SNS-matrix, we have

$$1 = (-1)^{c(L)}w(L) = (-1)^{1+n-k}w(\gamma) \cdot (-1)^{n-k} = (-1)w(\gamma).$$

Thus $w(\gamma) = -1$ for every cycle of length at least 2.

Now assume that the weight of each cycle equals -1. Let L be a linear subdigraph of $D^*(A)$ with k cycles (including cycles of length 1). Then $c(L) = k$ and $w(L) = (-1)^k$. Hence

$$(-1)^{c(L)}w(L) = (-1)^k \cdot (-1)^k = (-1)^{2k} = 1.$$

Hence A is an SNS-matrix. □

Example 9.3.9 Let

$$A = \begin{bmatrix} -1 & 1 & 0 & 0 \\ 0 & -1 & 1 & 1 \\ -1 & -1 & -1 & 1 \\ -1 & 0 & 0 & -1 \end{bmatrix}.$$

Each entry on the main diagonal of the matrix A equals -1, and hence Theorem 9.3.8 applies. The digraph $D(A)$ is pictured in Figure 9.1. One easily checks that each cycle of the digraph $D^*(A)$

has weight -1. Thus A is an SNS-matrix and its determinant equals the permanent of the matrix

$$\begin{bmatrix} 1 & 1 & 0 & 0 \\ 0 & 1 & 1 & 1 \\ 1 & 1 & 1 & 1 \\ 1 & 0 & 0 & 1 \end{bmatrix}.$$

The common value is 5.

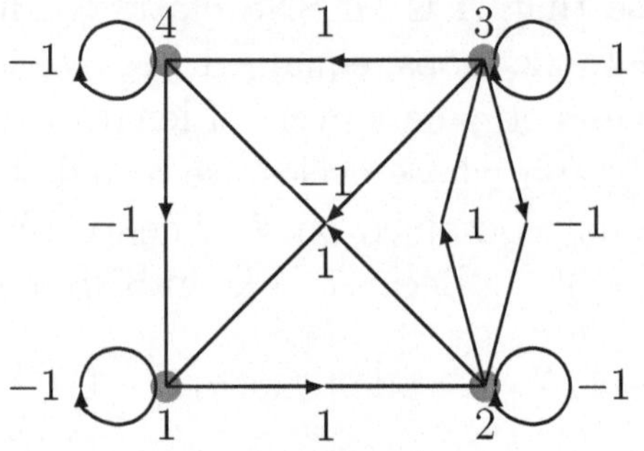

Figure 9.1

It is possible that a $(0, 1)$-matrix (or $(0, -1)$-matrix) be an SNS-matrix. Of course I_n and $-I_n$ are SNS-matrices, as are P and $-P$ for every permutation matrix P. A nontrivial example is the circulant

$$\begin{bmatrix} 1 & 1 & 0 & 1 & 0 & 0 & 0 \\ 0 & 1 & 1 & 0 & 1 & 0 & 0 \\ 0 & 0 & 1 & 1 & 0 & 1 & 0 \\ 0 & 0 & 0 & 1 & 1 & 0 & 1 \\ 1 & 0 & 0 & 0 & 1 & 1 & 0 \\ 0 & 1 & 0 & 0 & 0 & 1 & 1 \\ 1 & 0 & 1 & 0 & 0 & 0 & 1 \end{bmatrix}$$

whose permanent and determinant both equal 24. Replacing the 1's in this matrix by any numbers whatsoever results in a matrix whose determinant equals its permanent.

The permanent is an important and fundamental combinatorial function for which we have given a graphical interpretation. It motivated for us the important notion of an SNS-matrix. More about SNS-matrices and related topics can be found in [10].

9.4 Exercises

1. Let A be a matrix of order m, and let B be a matrix of order n. Let the eigenvalues of A be $\lambda_1, \lambda_2, \ldots, \lambda_m$, and let the eigenvalues of B be $\mu_1, \mu_2, \ldots, \mu_n$. Prove that the eigenvalues of $aA \otimes I_n + bI_m \otimes B$ are $a\lambda_i + b\mu_j$ $(1 \leq i \leq m, 1 \leq j \leq n)$.

2. Let A and B be as in Exercise 1. Prove that A and B do not have a common eigenvalue (a number that is an eigenvalue of both A and B) if and only if

$$\det(A \otimes I_n - I_m \otimes B) \neq 0.$$

3. Determine the Geršgorin region for each of the following matrices:

   (a) $A = \begin{bmatrix} 1 & 1 \\ 1 & -1 \end{bmatrix}$.

   (b) $A = \begin{bmatrix} 1 & 0 & 1 \\ 1 & -3 & 1 \\ 2 & 0 & 6 \end{bmatrix}$.

   (c) $A = \begin{bmatrix} 1+i & 2 & -1 \\ 3 & 2 & 1 \\ 2 & 1-i & 2-i \end{bmatrix}$.

4. Let $A = [a_{ij}]$ be a matrix of order n, and let D be a diagonal matrix of order n with positive diagonal entries $d_1, d_2, \ldots, d_n$. Apply Geršgorin's theorem to $D^{-1}AD$ to obtain an inclusion region for the eigenvalues of A.

5. Prove Theorem 9.2.5.

6. Prove Corollary 9.2.9.

7. Verify the properties of the permanent in Lemma 9.3.1.

8. Find a formula for the permanent of $A = (a-b)I_n + bJ_n$ in terms of the derangement numbers. (Here, as before, J_n denotes the matrix of order n each of whose entries equals 1.)

9. Compute the permanent of the *Hessenberg matrix* $H = [h_{ij}]$ of order n defined by

$$h_{ij} = \begin{cases} 1 & \text{if } i \leq j+1, \\ 0 & \text{otherwise.} \end{cases}$$

Thus H_4 is the matrix

$$\begin{bmatrix} 1 & 1 & 0 & 0 \\ 1 & 1 & 1 & 0 \\ 1 & 1 & 1 & 1 \\ 1 & 1 & 1 & 1 \end{bmatrix}.$$

10. Use Theorem 9.3.8 to determine whether or not the matrix

$$\begin{bmatrix} 1 & 0 & 1 & 1 & 0 & 0 \\ 1 & 1 & 1 & 0 & 0 & 0 \\ 0 & 1 & 1 & 0 & 0 & 1 \\ 0 & 1 & 0 & 1 & 0 & 0 \\ 1 & 0 & 0 & 0 & 1 & 0 \\ 0 & 0 & 0 & 0 & 1 & 1 \end{bmatrix}$$

is an SNS-matrix.

11. For each of the following matrices, show how to affix minus signs to some of the 1's so that the matrix becomes an SNS-matrix:

$$\begin{bmatrix} 0 & 1 & 1 & 0 \\ 1 & 0 & 1 & 1 \\ 1 & 1 & 0 & 1 \\ 1 & 1 & 1 & 0 \end{bmatrix}, \quad \begin{bmatrix} 1 & 1 & 0 & 1 \\ 1 & 1 & 1 & 0 \\ 0 & 1 & 1 & 1 \\ 1 & 0 & 1 & 1 \end{bmatrix}.$$

Chapter 10

Applications

This chapter is intended for those for whom mathematics is primarily a tool to describe and understand phenomena in other scientific disciplines. Because the applications of matrices and graphs in science are so numerous, we can present only a few selected examples. We shall focus on several topics where it is possible or even necessary to use the combinatorial approach developed in this book.

The three sections of this chapter describe some applications in electrical engineering, physics, and chemistry. These sections should not be considered as introductions or as surveys of these fields. In each section we assume a certain familiarity with the problems considered, with only very short explanations of background and specific terminology[1] given. However, we shall always give references to relevant general books where the interested reader can find more information. We also assume that the reader is acquainted with the fundamentals of basic mathematical analysis, in particular with ordinary and partial differential equations.

[1]Sometimes the terminology of a specific field conflicts with the terminology used in other chapters of this book.

10.1 Electrical Engineering: Flow Graphs

Applications of matrix theory in electrical engineering are numerous and varied. The matrices that appear in this area are usually sparse, i.e., contain a lot of zero entries. This fact justifies great popularity of graph-theoretical, i.e., combinatorial, methods in matrix theory among electrical engineers.

Electrical engineers have developed a series of methods for solving systems of linear algebraic equations, which appear in the theory of electrical circuits, control theory, and other areas. These methods use flow graphs (Coates [13], [26], [15]), signal flow graphs (Mason [59], [60], [80]) and Chan graphs (Chan and Mai [16], [14]); the first two graphs are described in Chapter 6. For more recent treatments, as well as for backgrounds, see, for example, [57] and [28].

It is noteworthy that the mentioned graphs (and especially, Mason's signal flow graphs) give a better insight into the physical system being described than the corresponding system of equations does. The signal flow graph technique is very effective and therefore popular among engineers. It was, in fact, first developed during the Second World War as an aid in designing weapon control systems by Shannon [73] but remained unknown to the public for many years.

One of the basic problems in the theory of electrical circuits[2] is to determine currents in all branches of a given electrical circuit when the voltages of electrical generators are given. For this purpose one uses the Kirchhoff Voltage Law (KVL), which says

---

[2]An electrical circuit is an interconnection of some two terminal components called *branches*. Branches are connected by their terminal points, called *nodes*. Electrical circuits have graphs as a natural mathematical model. Nodes become vertices and branches become edges of the associated graph. Depending on the problem, the associated graph can be undirected or directed. In the later case, an orientation is (in an arbitrary way) associated with each branch and (the same orientation) with the corresponding edge.

that the algebraic sum of voltage drops around any loop[3] is equal to zero. Applying KVL to several loops we get a system of linear algebraic equations with loop currents as unknowns. Graph theory helps to find a maximal set of *independent loops* ensuring that the obtained system of equations will consist of independent equations sufficent to determine all currents.

Example 10.1.1 We determine the current in the branch with resistance R_5 in the circuit of Figure 10.1.

We number the five loops in Figure 10.1 from 1 to 5 from left to right and orient them in a counterclockwise fashion. The equations for loop currents are

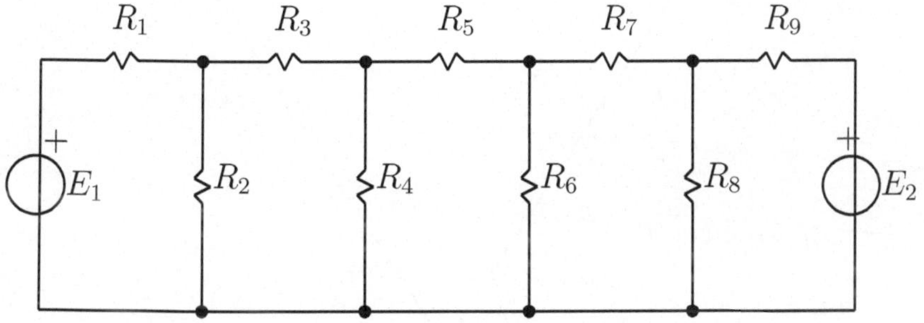

Figure 10.1

$$\begin{array}{llllll}
R_{11}I_1 & -R_{12}I_2 & & & & = -E_1, \\
-R_{21}I_1 & +R_{22}I_2 & -R_{23}I_3 & & & = 0, \\
 & -R_{32}I_2 & +R_{33}I_3 & -R_{34}I_4 & & = 0, \\
 & & -R_{43}I_3 & +R_{44}I_4 & -R_{45}I_5 & = 0, \\
 & & & -R_{54}I_4 & +R_{55}I_5 & = E_2,
\end{array}$$

where

$$R_{11} = R_1 + R_2, \quad R_{22} = R_2 + R_3 + R_4, \quad R_{33} = R_4 + R_5 + R_6,$$

$$R_{44} = R_6 + R_7 + R_8, \quad R_{55} = R_8 + R_9,$$

$$R_{12} = R_{21} = R_2, \; R_{23} = R_{32} = R_4,$$

[3]Here a loop means a subgraph that is reduced to a cycle if we neglect orientations of edges.

$$R_{34} = R_{43} = R_6, \; R_{45} = R_{54} = R_8.$$

Instead of explicitly writing this system of equations, we can immediately construct the corresponding Coates digraph (routine, after some practice). In our case the Coates digraph is given in Figure 10.2.

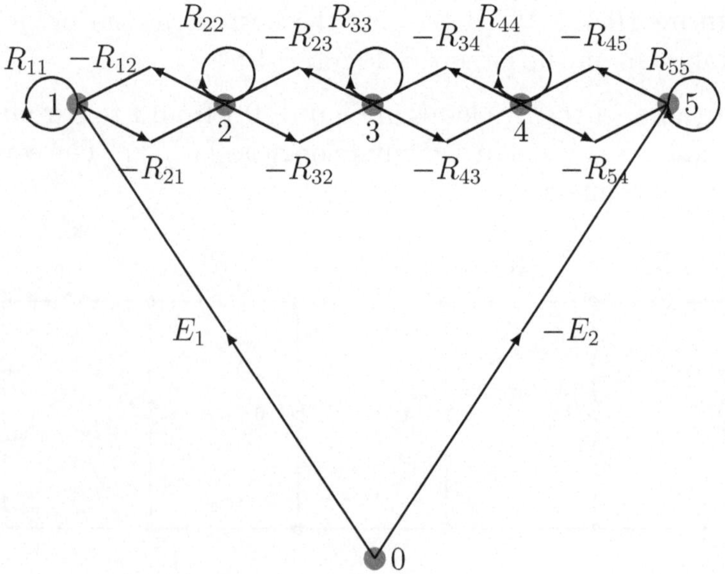

Figure 10.2

The current I_3 through R_5 corresponds to vertex 3 and can be immediately obtained by using Coates formula (6.15):

$$I_3 = \frac{N}{D},$$

where

$$N = E_1 R_{21} R_{32} R_{44} R_{55} - E_1 R_{21} R_{32} R_{45} R_{54} +$$
$$E_2 R_{45} R_{34} R_{12} R_{21} - E_2 R_{45} R_{34} R_{11} R_{22}$$

and

$$D = R_{12} R_{21} R_{33} R_{44} R_{55} - R_{12} R_{21} R_{34} R_{43} R_{55} - R_{12} R_{21} R_{33} R_{45} R_{54} +$$

$$R_{11} R_{23} R_{32} R_{44} R_{55} - R_{11} R_{23} R_{32} R_{45} R_{54} + R_{11} R_{22} R_{34} R_{43} R_{55} +$$

$$R_{11}R_{22}R_{33}R_{45}R_{54} - R_{11}R_{22}R_{33}R_{44}R_{55}.$$

□

In control theory, where systems and signals are the main objects, we usually encounter mathematical models that reduce to systems of ordinary linear diferential equations with constant coefficients. If to such a system we apply the Laplace transformation, we get a system of linear algebraic equations that is to be solved. Instead of time dependent functions, which represent voltages, currents, etc. (signals in general), the equations now contain the Laplace transforms of these functions. Matrices of these systems are typically sparse, and again we are in a position to apply techniques of combinatorial matrix theory, in particular signal and signal flow graphs. As their name indicates, signal flow graphs are specially designed to visualize and enable an easy analysis of the signal flow through complex systems of control theory. More information can be found, for example, in [5] and [28].

Example 10.1.2 For the two-terminal network of Figure 10.3, determine the transfer function $G(s)$ defined as the ratio of the Laplace transforms of the output $U_4(s)$ and imput signal $U_1(s)$.

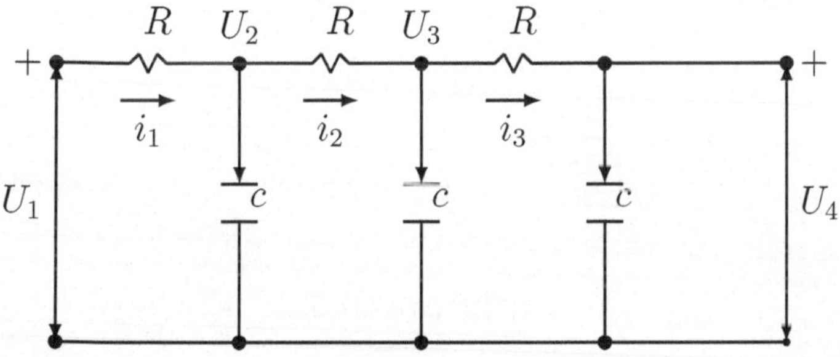

Figure 10.3

We have the following equations:

$$I_1(s) = \frac{1}{R}(U_1(s) - U_2(s)), \quad U_2(s) = \frac{1}{Cs}(I_1(s) - I_2(s)),$$

$$I_2(s) = \frac{1}{R}(U_2(s) - U_3(s)), \quad U_3(s) = \frac{1}{Cs}(I_2(s) - I_3(s)),$$

$$I_3(s) = \frac{1}{R}(U_3(s) - U_4(s)), \quad U_4(s) = \frac{1}{Cs}I_3(s).$$

The corresponding signal flow graph is given in Figure 10.4.

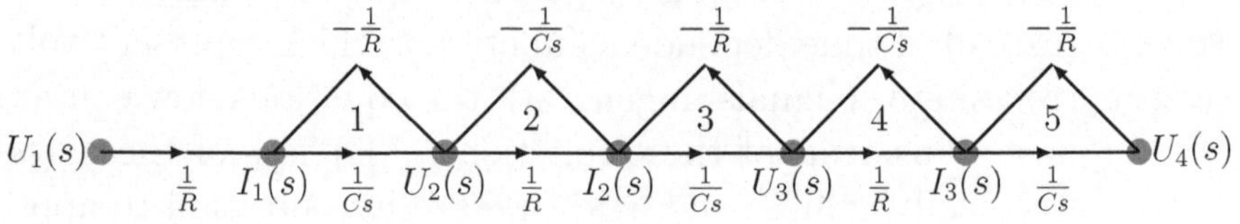

Figure 10.4

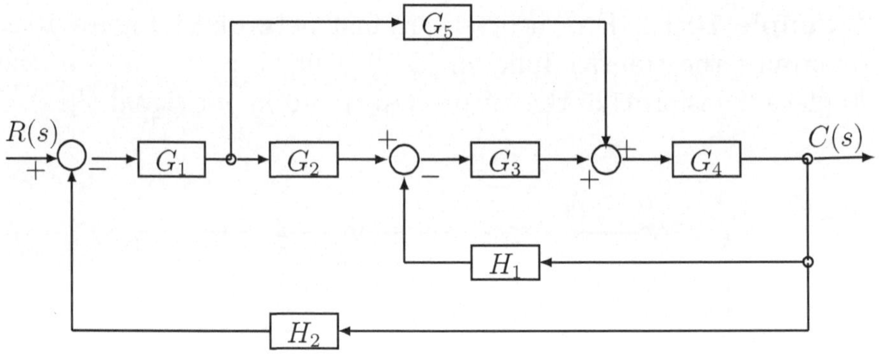

Figure 10.5

There exists just one path from vertex $U_1(s)$ to vertex $U_4(s)$. Its weight is $p_1 = \frac{1}{R^3C^3s^3}$. All cycles touch this path, and for the corresponding determinant we get $\Delta_1 = 1$.

There are five cycles, denoted by 1,2,3,4,5, in Figure 10.4, each of which has weight $-\frac{1}{RCs}$. There are exactly six pairs of cycles that do not touch each other, namely, cycles 1 and 3, 1 and 4, 1 and 5, 2 and 4, 2 and 5, and 3 and 5. There is only one triple 1,

3, and 5 of mutually nontouching cycles. Four or more cycles that mutually do not touch each other do not exist. Using Mason's formula we get

$$G(s) = \frac{\frac{1}{R^3C^3s^3}}{1-(-5\frac{1}{RCs})+6\frac{1}{R^2C^2s^2} - (-\frac{1}{R^3C^3s^3})}$$

$$= \frac{1}{\tau^3s^3+5\tau^2s^2+6\tau s+1}.$$

□

Example 10.1.3 For the system in Figure 10.5 we immediately get the corresponding signal flow graph in Figure 10.6.

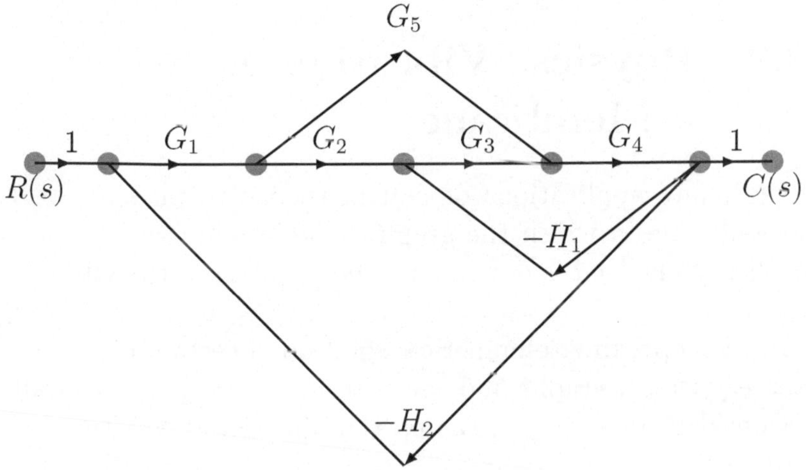

Figure 10.6

There are two paths from vertex $R(s)$ to vertex $C(s)$. They have weights

$$p_1 = G_1G_2G_3G_4, \quad p_2 = G_1G_5G_4.$$

The determinants of the corresponding subdigraphs are $\Delta_1 = \Delta_2 = 1$. The three cycles in the digraph do touch each other. The transfer function reads

$$\frac{C(s)}{R(s)} = \frac{G_1G_2G_3G_4 + G_1G_4G_5}{1+G_3G_4H_1+G_1G_2G_3G_4H_2+G_1G_4G_5H_2}.$$

□

The last example demonstrates some useful features of the signal flow graph technique: the variables in the equations of the system represented by a signal flow graph represent real signals in a technical system! The equation corresponding to the ith vertex of a signal flow graph (see (6.16)) says that the signal at the vertex i is equal to the sum of signals from other vertices via incoming edges multiplied by weights[4] of the edges and thus corresponds to the physical reality. In this way, a signal flow graph is, in fact, a simplified block diagram of a technical system (compare the block diagram of Figure 10.5 with the corresponding signal flow graph in Figure 10.6).

10.2 Physics: Vibration of a Membrane

Of the many applications of matrix theory to physics, we consider only one in which the graph-theoretic approach to the matrix theory involved is dominant—the problem of the vibration of a membrane.

In the approximate numerical solution of certain partial differential equations, graphs and their spectra arise quite naturally.

Consider, for example, the partial differential equation

$$\frac{\partial^2 z}{\partial x^2} + \frac{\partial^2 z}{\partial y^2} + \lambda z = 0 \tag{10.1}$$

(or $Az + \lambda z = 0$; where $A = \frac{\partial^2 z}{\partial x^2} + \frac{\partial^2 z}{\partial y^2}$ is the Laplace operator). Here the unknown function $z = z(x, y)$ is subject to the boundary condition $z(x, y) = 0$ on a simple closed curve ℓ' lying in the xy-plane. It is known that equation (10.1) has a solution only for an infinite sequence $\lambda_1 \leq \lambda_2 \leq \ldots \leq \lambda_n \leq \ldots$ of (discrete) values of λ, which are called the *eigenvalues* of the equation. The sequence of eigenvalues is called the *spectrum* of the equation, and the solutions of (10.1) are the corresponding *eigenfunctions*.

[4]In applications the weights of edges are called *transmittances*.

In an approximate determination of z we consider the values only for a set of points (x_i, y_i) that form a regular lattice (square, triangular, or hexagonal) in the xy-plane. A corresponding (infinite) graph can be associated, in a straightforward and natural way, with this lattice. Points (x_i, y_i) are the vertices of the graph and the edges connect pairs of points of minimal distance. The points (respectively, vertices) lying in the interior of Γ are called *internal points* (respectively, *internal vertices*), and the other points (respectively, vertices) of the lattice are called *external*. Let $z_i = z(x_i, y_i)$. Because of the boundary condition, we can take $z_i = 0$ for all external points.

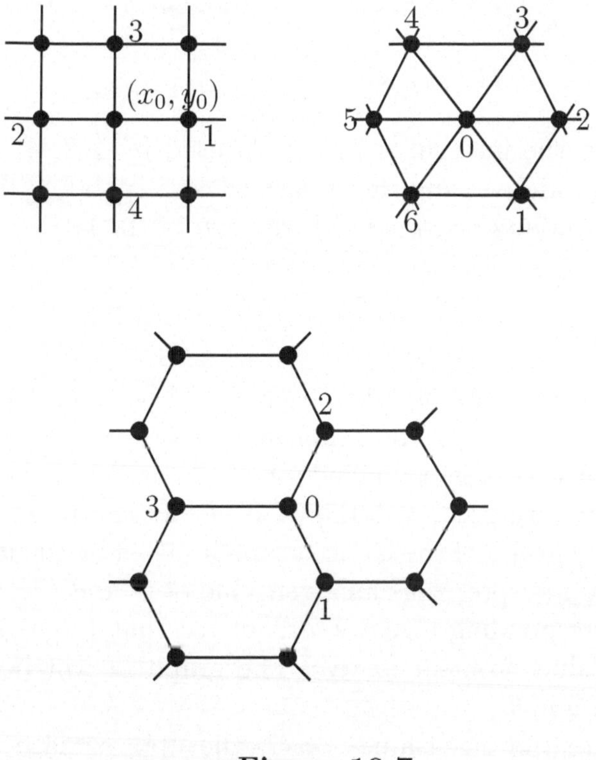

Figure 10.7

In the case of a square lattice (Figure 10.7), let $z_0 = z(x_0, y_0)$, (x_0, y_0) being a fixed point of the lattice, and let $z_1 = z(x_0+h, y_0)$, $z_2 = z(x_0 - h, y_0)$, $z_3 = z(x_0, y_0+h)$, and $z_4 = z(x_0, y_0 - h)$ be the values of z for the neighboring points (we assume that the points of the lattice lie on lines that are parallel with the coordinate axes and that the distance between any two neighboring points is h). The value of $\frac{\partial^2 z}{\partial x^2} + \frac{\partial^2 z}{\partial y^2}$ at the point (x_0, y_0) can, as usual, be approximated by

$$\frac{1}{h^2}(z_1 + z_2 + z_3 + z_4 - 4z_0).$$

Equation (10.1) then becomes

$$\frac{1}{h^2}(z_1 + z_2 + z_3 + z_4 - 4z_0) + \lambda z_0 = 0, \textit{ or}$$

$$(4 - \lambda h^2)z_0 = z_1 + z_2 + z_3 + z_4. \tag{10.2}$$

Now let the internal points be labeled by $1, 2, \ldots, n$. Taking $\nu = 4 - \lambda h^2$ and writing the equations corresponding to (10.2) for all internal points (x_i, y_i), $i = 1, 2, \ldots, n$ of the lattice, we obtain

$$\nu z_i = \sum_{j_i} z_{j_i} \quad (i = 1, 2, \ldots, n), \tag{10.3}$$

where the summation is taken over all indices j_i corresponding to internal points (x_{j_i}, y_{j_i}) neighboring (x_i, y_i).

It is not necessary to include in the sum (10.3) those external points neighboring (x_i, y_i) if the value of z for this point is zero. Let G be the subgraph of the lattice graph induced by the internal vertices. If we interpret ν as an eigenvalue of G and $(z_1, z_2, \ldots, z_n)^T$ as the corresponding eigenvector, we see that (10.3) just defines the eigenvalue problem for G. The graph G will be called the *membrane graph*.

If ν_i are the eigenvalues of G, the approximate eigenvalues of equation (10.1) are given by $\lambda_i^* = \frac{4 - \nu_i}{h^2}$. The corresponding eigenvalues of G represent an approximate solution of (10.1). Note that the λ_i^* $(i = 1, 2, \ldots, n)$ do not necessarily represent approximate values for the first n eigenvalues $\lambda_1, \ldots, \lambda_n$ of (10.1), but for some eigenvalues $\lambda_{i_1}, \ldots, \lambda_{i_n}$.

The adjacency matrix A of the graph G is a sparse matrix. For an arbitrary large n, the number of nonzero entries in any row or in any column is not greater than the vertex degrees in the lattice (i.e., 4, 6 and 3 for the square, triangular and hexagonal lattices, respectively). Therefore, such matrices can be treated, at least in principle, by those methods described in Section 6.5. The digraphs $D(A)$ and $D^*(A)$ are identical because A is a symmetric matrix and a modification of them is just the membrane graph.

For the triangular and hexagonal lattices (see Figure 10.7), we have, respectively, the following approximate expressions for Az in the point (x_0, y_0):

$$\frac{2}{3h^2}(z_1 + z_2 + z_3 + z_4 + z_5 + z_6 - 6z_0),$$

$$\frac{4}{3h^2}(z_1 + z_2 + z_3 - 3z_0).$$

We again obtain (10.3), but now the connection between the eigenvalues of G and of (10.2) is given by $\lambda_i^* = \frac{2}{3}\frac{6-\nu_i}{h^2}$ and $\lambda_i^* = \frac{4}{3}\frac{3-\nu_i}{h^2}$, respectively.

The procedure described for approximately solving a partial differential equation is often used in technical problems (see, for example, [18]). In this way the theory of graph spectra can be very useful in practical calculations.

The most interesting problem that can be treated by such a procedure is that of *membrane vibration*. There are some other problems of the same kind, for example, air oscillations in space, etc. (see [19], [18], [51], and [69]). These problems motivated the authors of [19] to consider graph spectra.

If a vibrating membrane Ω is held fixed along its boundary Γ, its displacement $F(x, y, t)$ in the direction orthogonal to its plane is a function of the coordinates x, y and time t and satisfies the wave equation

$$\frac{\partial^2 F}{\partial t^2} = c^2\left(\frac{\partial^2 F}{\partial x^2} + \frac{\partial^2 F}{\partial y^2}\right), \tag{10.4}$$

where c is a constant depending on the physical properties of the membrane and of the tension under which the membrane is held.

The solutions of the form $F(x, y, t) = z(x, y)e^{i\omega t}$ are of particular interest. If we substitute this expression in (10.4), we obtain

$$-\omega^2 z(x, y) = c^2 \left(\frac{\partial^2 z(x, y)}{\partial x^2} + \frac{\partial^2 z(x, y)}{\partial y^2} \right). \tag{10.5}$$

Setting $\lambda = \frac{\omega^2}{c^2}$ reduces (10.5) to (10.1).

The representation of a membrane by a graph is by no means a mathematical abstraction. Equation (10.1) describes the vibration of a membrane. The membrane is represented by a continuous model. If the membrane is described by a discrete model as given below, we arrive at the system (10.3) obtained in the approximate solution.

According to the discrete model, the membrane consists of a set of atoms that in the equilibrium state, lie on the vertices of a regular lattice graph embedded in a plane. Each atom acts on its neighboring atoms by elastic forces. We assume that all atoms have the same mass and that elastic forces are of the same intensity for all neighboring pairs of atoms. If $z_i(t)$ and $z_j(t)$ are displacements of neighboring atoms i and j at time t, the elastic force tending to reduce the relative displacement between these atoms is

$$F_{ij} = -K\left(z_i(t) - z_j(t)\right),$$

where K is a constant characteristic of the elastic properties of the membrane.

The equation of motion of the kth atom is

$$m\frac{\mathrm{d}^2 z_k(t)}{\mathrm{d}t^2} = -K \sum_{j_k} \left(z_k(t) - z_{j_k}(t)\right), \tag{10.6}$$

where m is the mass of an atom and where the summation is taken over the nearest neighbors j_k of the kth atom. For a vertex j of the lattice graph in which there is no atom of the membrane, we have $z_j(t) = 0$ (as before, such vertices are called *external*).

We can again consider pure harmonic oscillations and take $z_k(t) = z_k e^{i\omega t}$ (where $i = \sqrt{-1}$). If we insert this expression into (10.6) for each atom k, then we again obtain the graph eigenvalue problem (10.3). Thus, a solution of the discrete model is equivalent to an approximate solution of the continuous model. (Of course, in some cases the whole thing can be considered the other way around; the continuous model could give an approximate solution for the discrete model).

We conclude this section with the following observation. If the problems of continuous mathematics (analysis) are to be solved by means of computers, they must be approximated by the corresponding discrete models since computers operate with discrete actions. They change their state in discrete moments and an inner state of a computer is determined by the states of a finite number of computer cells where the number of states of any cell is finite. Therefore numerical mathematics represents a link (a union of sorts) of the continuous and discrete.

10.3 Chemistry: Unsaturated Hydrocarbons

In this section[5] we present a specific chemical application of matrix theory and graph theory. The applications of matrices and graphs in chemistry (especially in physical and theoretical chemistry) are so numerous that it is impossible to give any reasonable survey in a limited space. The interested reader may consult [50]. Our discussion is in four parts.

Hückel Molecular Orbital Theory

One of the basic goals of quantum chemistry is to describe the electronic structure of molecules. This can be done by solving the Schrödinger equation

$$\hat{H}\Psi_j = E_j\Psi_j, \tag{10.7}$$

where $\hat{H}$ is the Hamiltonian operator (or energy operator), Ψ_j is the wave function of the system under consideration, and E_j is the energy of the system. The subscript j indicates that in the general case a Schrödinger equation has more than one solution.

[5]This section is based on a chapter of [23] that was written by I. Gutman.

The wave function Ψ_j fully describes the jth state of the system whose Hamiltonian operator is $\hat{H}$.

If the wave function describes the state of an electron in a molecule, then it is called a *molecular orbital.* The physical meaning of a molecular orbital $\Psi = \Psi(x, y, z)$ is that $|\Psi(x, y, z)|^2 \mathrm{d}x\,\mathrm{d}y\,\mathrm{d}z$ is the probability of finding the pertinent electron in the volume element $\mathrm{d}V = \mathrm{d}x\,\mathrm{d}y\,\mathrm{d}z$ at the point with the space coordinates x, y, z.

The Hamiltonian operator requires (among other operations) the calculation of the second partial derivatives with respect to the space coordinates x, y, and z. Thus, the Schrödinger equation is a second order partial differential equation. Under certain conditions (which we will not specify here), the differential equation (10.7) can be transformed into matrix form:

$$H\Psi_j = E_j\Psi_j, \tag{10.8}$$

where now H is a Hamiltonian matrix and Ψ_j is the wave function in vector form. From (10.8) it is evident that Ψ_j is the eigenvector and E_j is the eigenvalue of the matrix H.

In order to solve the Schrödinger equations (10.7) and (10.8) for complicated many-electron molecular systems, various approximations are used. In the pioneering days of quantum chemistry (in the 1930s and 1940s) an approximate method for describing the state of single electrons in conjugated hydrocarbons was developed, known under the name *Hückel molecular orbital theory.*[6]

Within the framework of the Hückel method, the Hamiltonian matrix $H = [h_{ij}]$ is a square matrix of order n, where n is the

[6] *Hydrocarbons* are chemical compounds composed of only two elements—carbon (C) and hydrogen (H). A hydrocarbon is *saturated* if its molecules possess only single bonds. If in a molecule there are also multiple bonds, then the hydrocarbon is *unsaturated.* An important class of unsaturated hydrocarbons is the *conjugated* hydrocarbons, each of whose carbon atoms participates in exactly one double bond. We assume that in a hydrocarbon molecule all carbon atoms have valency 4 and all hydrogen atoms have valency 1.

The Hückel *graph* [42] is used for an abbreviated representation of conjugated hydrocarbons. Its vertices represent only the carbon atoms, and all its edges are simple (irrespective of whether the corresponding chemical bonds are single or double). The vertices of a Hückel graph may be of degree 1,2, or 3.

number of carbon atoms in the molecule. Let these carbon atoms be labeled by $1, 2, \ldots, n$. Then the matrix elements h_{rs} are given by

$$h_{rs} = \begin{cases} \alpha & \text{if } r = 1, 2, \ldots, n \\ \beta & \text{if } r \neq s \text{ and the atoms } r \text{ and } s \text{ are chemically bonded} \\ 0 & \text{if } r \neq s \text{ and no chemical bond between the atoms} \\ & r \text{ and } s \text{ exists.} \end{cases} \tag{10.9}$$

The parameters α and β are called the *Coulomb* and the *resonance integral*; in Hückel theory these are assumed to be constants. The approximations imposed by the relations (10.9) are severe. Therefore it is surprising that the results of the Hückel theory are (at least sometimes) in good agreement with both experimental findings and other, more advanced, theoretical approaches [67].

For example, for the hydrocarbon styrene (I, Figure 10.8) the Hückel–Hamiltonian matrix has the form

$$H = \begin{bmatrix} \alpha & \beta & 0 & 0 & 0 & 0 & 0 & 0 \\ \beta & \alpha & \beta & 0 & 0 & 0 & 0 & 0 \\ 0 & \beta & \alpha & \beta & 0 & 0 & 0 & \beta \\ 0 & 0 & \beta & \alpha & \beta & 0 & 0 & 0 \\ 0 & 0 & 0 & \beta & \alpha & \beta & 0 & 0 \\ 0 & 0 & 0 & 0 & \beta & \alpha & \beta & 0 \\ 0 & 0 & 0 & 0 & 0 & \beta & \alpha & \beta \\ 0 & 0 & \beta & 0 & 0 & 0 & \beta & \alpha \end{bmatrix}.$$

Keeping in mind relations (10.9), we see that the Hückel–Hamiltonian matrix can be presented as

$$H = \alpha I_n + \beta A, \tag{10.10}$$

where A is a symmetric matrix whose diagonal elements equal 0 and whose off-diagonal elements equal 1 or 0, depending on whether the corresponding atoms are connected or not. In fact, $A = A_H$ is just the adjacency matrix of the Hückel graph. Graph II in Figure 10.8 is the Hückel graph of styrene I. Equation (10.10) immediately gives the following result.

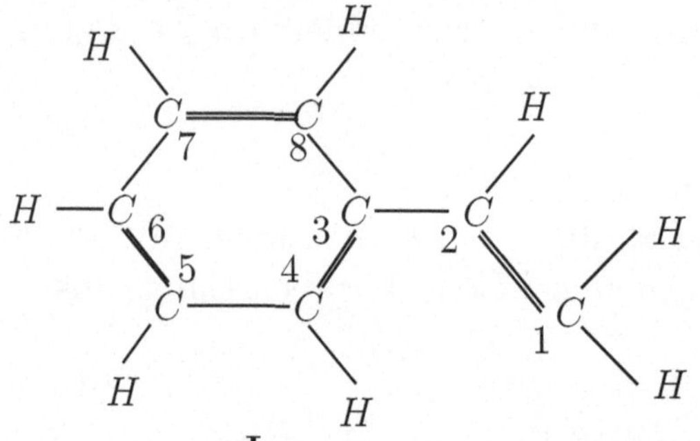

I

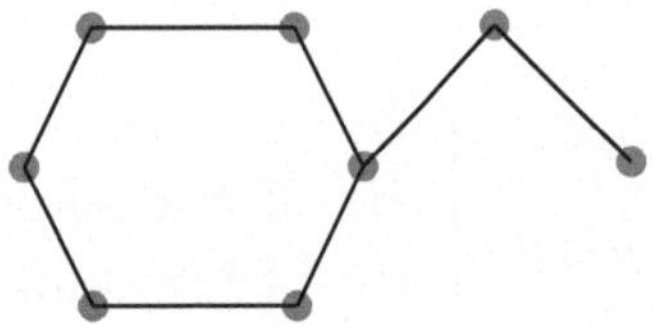

II

Figure 10.8

Theorem 10.3.1 *If λ is an eigenvalue and z is an eigenvector of the matrix A, then $\alpha + \beta\lambda$ is an eigenvalue and z is an eigenvector of the matrix H.* □

From this theorem it follows that the Hückel molecular orbitals Ψ_j coincide with the eigenvectors z_j of the adjacency matrix of the Hückel graph, that is, $\Psi_j = z_j$. The eigenvalues λ_j of the matrix A_H and the energies E_j of the corresponding electrons are related simply as

$$E_j = \alpha + \beta\lambda_j.$$

There are exactly n different molecular orbitals, namely, the z_j for $j = 1, 2, \ldots, n$.

This important conclusion shows that there is a deep and far-reaching relation between the Hückel molecular orbital theory and graph spectral theory. The Hückel theory provides an important field of application of the graph spectra.

For more information on Hückel theory the interested reader can consult, for example, [2], [21], [27], [40], [36], [77], and [24].

Two Examples: Linear Polyenes and Annulenes

We are now going to determine the characteristic polynomials and spectra of two important Hückel graphs, namely, those associated with linear polyenes and annulenes, as given in Figure 10.9 for $n = 8$. The Hückel graphs of these compounds are given in

Figure 10.10 and are, respectively, the path P_n with n vertices and the circuit C_n of length n.

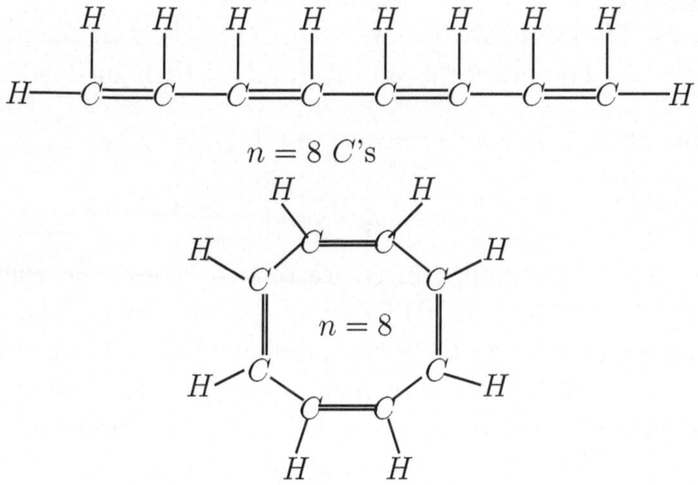

Figure 10.9

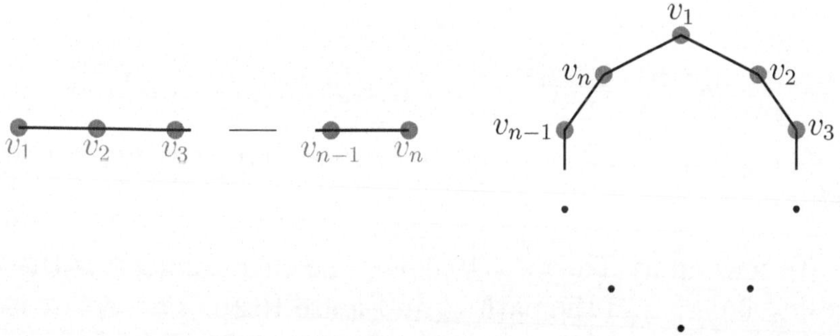

Figure 10.10

We prove first some general results [41], [42]. In Section 8.5, the characteristic polynomial $p(W, \lambda)$ of a forest W with n vertices was shown to satisfy the formula

$$p(W, \lambda) = \sum_{k=1}^{\left[\frac{n}{2}\right]} (-1)^k m(W, k) \lambda^{n-2k}, \qquad (10.11)$$

where $m(W,k)$ is the number of k-matchings in W.

Let G be a graph with n vertices $v_1, v_2, \ldots, v_n$ and let e_{rs} be an arbitrary edge of G connecting the vertices v_r and v_s. The graphs $G - e_{rs}$ and $G - v_r - v_s$ are obtained from G by deleting, respectively, the edge e_{rs} and vertices v_r and v_s (and all their incident edges).

Lemma 10.3.2 *For an arbitrary graph G, we have*

$$m(G,k) = m(G - e_{rs}, k) + m(G - v_r - v_s, k-1). \qquad (10.12)$$

Proof. The k-matchings in G are of two types: the edge e_{rs} is (i) in the k-matching or (ii) is not in the matching. The number of k-matchings of type (i) is the number of $(k-1)$-matchings of $G - v_r - v_s$, and thus equals $m(G - v_r - v_s, k-1)$. The number of k-matchings of type (ii) is the number of k-matchings of $G - e_{rs}$, and thus equals $m(G - e_{rs}, k)$. The lemma now follows. □

Combining equations (10.11) and (10.12), we obtain the following result.

Theorem 10.3.3 *The characteristic polynomial of a forest W satisfies the recurrence relation*

$$p(W,\lambda) = p(W - e_{rs}, \lambda) + p(W - v_r - v_s, \lambda)$$

where e_{rs} is an arbitrary edge of W that connects the vertices v_r and v_s. □

We now apply Theorem 10.3.3 to the edge connecting the vertices v_n and v_{n-1} of the path P_n in Figure 10.10. $P_n - e_{n,n-1}$ is the graph with connected components P_{n-1} and P_1. Therefore

$$p(P_n - e_{n,n-1}, \lambda) = p(P_1, \lambda)p(P_{n-1}, \lambda) = \lambda p(P_{n-1}, \lambda).$$

In addition, $P_n - v_n - v_{n-1}$ is the path with $n-2$ vertices. Thus from Theorem 10.3.3 we obtain

$$p(P_n, \lambda) = \lambda p(P_{n-1}, \lambda) - p(P_{n-2}, \lambda), \qquad (10.13)$$

from which it is easy to recursively evaluate the polynomials $p(P_n, \lambda)$ starting with $p(P_1, \lambda) = \lambda$ and $p(P_2, \lambda) = \lambda^2 - 1$.

In the theory of special functions, the Chebyshev functions $T_n(\lambda)$ of the first kind and the Chebyshev function $U_n(\lambda)$ of the second are investigated. These are the two independent *particular* solutions of the differential equation

$$\frac{\mathrm{d}^2 y}{\mathrm{d}\, t^2} + n^2 y = 0,$$

where $t = \cos\lambda$. The Chebyshev functions satisfy the recurrence relations

$$T_n(\lambda) = 2\lambda T_{n-1}(\lambda) - T_{n-2}(\lambda),$$
$$U_n(\lambda) = 2\lambda U_{n-1}(\lambda) - U_{n-2}(\lambda),$$

whose forms are quite similar to that of equation (10.13).

Knowing that $U_2(\lambda) = 2\lambda\sqrt{1-\lambda^2}$ and $U_3(\lambda) = (4\lambda^2 - 1)\sqrt{1-\lambda^2}$, it is easy to verify that

$$p(P_n, 2\lambda)\sqrt{1-\lambda^2} = U_{n+1}(\lambda). \qquad (10.14)$$

This identity (10.14) is an example of the interesting connections that exist between graphs and special functions.

The general solution of the recurrence relation $f_n = af_{n-1} + bf_{n-2}$ is $f_n = Ax_1^n + Bx_2^n$, where A and B are constants and x_1 and x_2 are the roots of the equation $x^2 = ax + b$. We apply this fact to (10.13). The roots of the equation $x^2 = \lambda x - 1$ are $x_{1,2} = \dfrac{\lambda \pm \sqrt{\lambda^2 - 4}}{2}$. After substituting $\lambda = 2\cos t$, we get

$$x_{1,2} = \cos t \pm \sqrt{\cos^2 t - 1} = \cos t \pm i \sin t.$$

Taking into account the Euler formula $\cos t \pm i \sin t = e^{\pm it}$, we further obtain

$$\begin{aligned} p(P_n, 2\cos t) &= A\left(e^{it}\right)^n + B\left(e^{-it}\right)^n \\ &= Ae^{int} + Be^{-int} \\ &= (A+B)\cos nt + i(A-B)\sin nt. \end{aligned}$$

The constants A and B are determined from the initial conditions:

$$p(P_1, 2\cos t) = 2\cos t \text{ and } p(P_2, 2\cos t) = 4\cos^2 t - 1.$$

Elementary calculation gives $A+B=1$ and $i(A-B)=\dfrac{\cos t}{\sin t}$, so that

$$\begin{aligned} p(P_n, 2\cos t) &= \cos nt + \cos t\,\frac{\sin nt}{\sin t} \\ &= \frac{(\cos nt \sin t + \sin nt \cos t)}{\sin t} \\ &= \frac{\sin(n+1)t}{\sin t}\,. \end{aligned}$$

Consequently, the characteristic polynomial of the path with n vertices has the following simple form:

$$p(P_n, \lambda) = \frac{\sin(n+1)t}{\sin t}, \tag{10.15}$$

with $\lambda = 2\cos t$. The spectrum of the path now follows immediately from the equation $p(P_n, \lambda) = 0$. The condition $\sin(n+1)t = 0$ implies $(n+1)t = \pi j$, that is,

$$\lambda_j = 2\cos\frac{\pi j}{n+1},\ j = 1, 2, \ldots, n\,.$$

We now calculate the characteristic polynomial and spectrum of the circuit C_n. The graph C_n is a connected graph with n vertices and n edges. Theorem 8.4.5 from Section 8.5 implies that this basic figure effects only the coefficient a_n and its contribution to a_n is equal to $(-1)^1 2^1 = -2$. All other basic figures are composed exclusively of graphs K_2. Therefore, the coefficients of the characteristic polynomial of the cycle C_n are determined as $a_j = b_j$ for $j = 1, 2, \ldots, n-1$ and $a_n = b_n - 2$, where

$$b_{2k} = (-1)^k p(C_n, k),\ k = 1, 2, \ldots,\ \text{and}$$

$$b_{2k-1} = 0, \quad k = 1, 2, \ldots.$$

This result can be formulated as

$$p(C_n, \lambda) = -2 + \sum_{k=0}^{\left[\frac{n}{2}\right]} (-1)^k p(C_n, 2k)\lambda^{n-2k}. \tag{10.16}$$

We now apply Lemma 10.3.2 to the edge e_{1n} connecting the vertices v_1 and v_n of the cycle C_n. It is easily seen that $C_n - e_{1n} = P_n$ whereas $C_n - v_1 - v_n = P_{n-2}$. Hence $m(C_n, k) = m(P_n, k) + m(P_{n-2}, k-1)$, which when substituted back into (10.16), gives

$$p(C_n, \lambda) = p(P_n, \lambda) - p(P_{n-2}, \lambda) - 2 \qquad (10.17)$$

Using (10.15), we have further

$$p(C_n, \lambda) = \frac{\sin(n+1)t}{\sin t} - \frac{\sin(n-1)t}{\sin t} - 2 = 2(\cos nt - 1), \quad (10.18)$$

where $\lambda = 2\cos t$. From (10.18) we can directly determine the spectrum of C_n. If $p(C_n, \lambda) = 0$, then $nt = 2\pi j$, and thus

$$\lambda_j = 2\cos\frac{2\pi j}{n}, \quad j = 1, 2, \ldots, n.$$

The Chebyshev functions of the first kind and the characteristic polynomial of the cycle are related by

$$P(C_n, 2\lambda) = 2T_n(\lambda) - 2.$$

Finally we note that the knowledge of the spectrum of the graphs P_n and C_n is of great importance in the quantum chemical descriptions of the electronic structure of linear polyenes and annulenes. In particular, it is important that the spectrum of the cycle C_n possesses (two) zeros if and only if $n = 4l$ $(l = 1, 2, \ldots)$. It will be explained in the next part of this section that this means that the annulenes with $4l$ carbon atoms have nonbonding molecular orbitals and that these compounds are chemically unstable.

Stability of a Molecule

We shall discuss here some problems and results of Hückel theory that can be formulated in a particularly simple way using graph-theoretical terminology.

A molecular orbital Ψ_j with the energy $E_j = \alpha + \beta\lambda_j$ is called *bonding* if $\lambda_j > 0$, *antibonding* if $\lambda_j < 0$, and *nonbonding* if $\lambda_j = 0$. The electrons in the bonding molecular orbitals strengthen the chemical bonds in the molecule, whereas the effect of the electrons in antibonding orbitals is just the opposite. The electrons whose state is described by non-bonding molecular orbitals play a less pronounced role in the creation of chemical bonds. However, within the framework of the Hückel theory, it can be demonstrated that conjugated hydrocarbons possessing nonbonding molecular orbitals are extremely unstable and chemically reactive. The origin of this phenomenon cannot be explained here.

In the theory of conjugated compounds it is quite important to establish which systems have nonbonding molecular orbitals. Evidently, the number of nonbonding molecular orbitals coincides with the multiplicity of 0 in the spectrum of the pertinent Hückel graph. Because $\det A = \prod_{j=1}^{n} \lambda_j$, 0 is in the spectrum of a graph if and only if $\det A = 0$.

A general solution of the problem of finding the multiplicity of 0 in the spectrum of a graph is not known, but a variety of partial results have been obtained. As an illustration we present the following two statements (see also [24], Section 8.1).

Theorem 10.3.4 *Assume the graph G has a vertex v_r of degree 1, where v_r is adjacent to the vertex v_s. Then the graphs G and $G - v_r - v_s$ have equal multiplicity of the number 0 in their spectra.*

Proof. Let $c = [c_1, c_2, \ldots, c_n]$ be an eigenvector of the adjacency matrix $A = [a_{ij}]$ of some graph on n vertices corresponding to the eigenvalue 0. Then $Ac = 0$ and so

$$\sum_{q=1}^{n} a_{pq} c_q = 0 \quad (p = 1, 2, \ldots, n).$$

Because $a_{pq} = 0$ when the vertices v_p and v_q are not adjacent and $a_{pq} = 1$ if these vertices are adjacent, we get

$$\sum_{q_p} c_{q_p} = 0 \quad (p = 1, 2, \ldots, n), \tag{10.19}$$

where the summation is over all q_p such that the vertex v_{q_p} is adjacent to v_p. Hence if c is an eigenvector of a graph for eigenvalue 0, the sum of the components of the vector c "around" each vertex v_p equals zero. Hence the number of linearly independent eigenvectors with eigenvalue 0, equivalently, the multiplicity of 0 in the spectrum of a graph, is equal to the number of independent components c_p in the system of equations (10.19).

In the graph G, let the vertex v_s be adjacent to the vertices $v_a, v_b, \ldots, v_f$ in addition to vertex v_r. Let the system (10.19) be satisfied for the graph $G - v_r - v_s$ so that the vertices $v_a, v_b, \ldots, v_f$ correspond to the components $c_a, c_b, \ldots, c_f$ of the vector c.

Now consider the graph G and the system (10.19). Then we have to add to the system (10.19) for $G - v_r - v_s$ two further equations: $c_s = 0$ and $c_a + c_b + \cdots + c_f + c_r = 0$. Because of the condition $c_s = 0$, all equations (10.19) that were valid for the graph $G - v_r - v_s$ also hold for G. Because $c_s = 0$ and $c_r = -(c_a + c_b + \cdots + c_f)$ are evidently not new independent variables, we see that the number of independent components of the vector c in the graphs $G - v_r - v_s$ is the same as that for G. □

Theorem 10.3.5 *Assume that the graph G has a path*

$$v_r, v_a, v_b, v_c, v_d, v_s, v_s$$

where the vertices v_a, v_b, v_c, v_d have degree equal to 2. Let the graph G' be obtained from G by deleting the vertices v_a, v_b, v_c, and v_d and introducing a new edge e_{rs} between the vertices v_r and v_s. Then G and G' have equal multiplicity of the number 0 in their spectra.

Proof. In order that the system (10.19) be satisfied for the graph G, the following equations (among others) must hold:

$$c_r + c_b = 0, \;\; c_a + c_c = 0, \;\; c_b + c_d = 0, \;\; c_c + c_s = 0.$$

Consequently, $c_a = c_s$ and $c_d = c_r$. Therefore, the deletion of v_a, v_b, v_c, and v_d and the simultaneous connection of v_r and v_s cause no change in the number of independent components of the vector c. □

Alternant Hydrocarbons and Their Graphs

In theoretical chemistry, a frequently used concept is that of *alternant hydrocarbons.* A conjugated hydrocarbon is said to be alternant if all its atoms can be simultaneously labeled by two labels (usually called star and circle), so that every atom labeled by a star has only neighbors labeled by a circle and viceversa. Hydrocarbons for which such a labeling is not possible are called *nonalternant.*

The labeling of the atoms of the molecule by stars and circles is equivalent to the coloring of the vertices of the molecular graph by two colors (as described in Section 1.1). Therefore alternant (respectively, nonalternant) hydrocarbons have bipartite (respectively, nonbipartite) molecular graphs.

In Section 8.3 it was proved that the spectrum of a bipartite graph is symmetric with respect to zero, that is, λ and $-\lambda$ are eigenvalues with equal multiplicity. In Hückel theory this result is interpreted to mean that the energy levels of the molecular orbitals are symmetrically distributed around the energy $E_0 = \alpha$. Consequently, the orbital with energy $E = \alpha + \beta\lambda$ is "paired" with an orbital with energy $E = \alpha - \beta\lambda$. This is the famous "Pairing theorem" that has a number of important consequences in quantum chemistry.

10.4 Exercises

1. Assume that the graph G has two vertices v_r and v_p of degree 1 that are adjacent to the same vertex v_s. Prove that 0 is an eigenvalue of G.

2. Let M be the size of the maximal matching in a tree T on n vertices. Show that the multiplicity of the eigenvalue 0 in the spectrum of T is equal to $n - 2M$.

3. Prove or disprove: The adjacency matrix of a tree T is regular if and only if for each vertex v, the forest $T - v$ has exactly one component with an odd number of vertices.

4. Find values of n for which all eigenvalues of the circuit C_n are integers.[7]

[7] Graphs whose spectra consist entirely of integers are called *integral graphs*.

Coda

As remarked in the preface, the graph-theoretical connections with matrix theory are numerous, and emphasizing them often leads to a clearer and deeper understanding of many of the concepts and results of matrix theory. The first systematic use of graphs with matrices seems to be by König [55]. We have introduced several (weighted) graphs that can be associated to a matrix—the König digraph and the (Coates) digraph of a matrix being the most prominent of them. The digraph $G(A)$ of the matrix A is called the König digraph, because König used the corresponding bipartite graph in his papers (see [56]). The digraph $D^*(A)$ is named the Coates digraph and formula (4.1) is called the Coates formula, because C.L. Coates introduced them in [13], although it is very hard to establish who first came to the idea of such a graphical interpretation of a determinant (see the discussion about this in Chapter 1 of [24]). F. Harary, referring to C.L. Coates, has proposed in [45] that this formula, in a somewhat changed form, could be taken as the definition of the determinant. Therefore, this definition could be called the Harary–Coates definition. However, it seems that Harary's suggestion from [45] was forgotten, and one of the authors of this book later independently came to the same idea and, starting from it, outlined the elementary theory of determinants [22]. A similar development appeared a little bit later and again independently [37]. These two digraphs have been used to illuminate the basic algebraic properties of matrices, including matrix multiplication, determinants, inverses of matrices, cofactors, and Cramer's formula, and including solutions of linear systems of equations.

Consideration of the digraph often leads to an easier description of many matrix properties. It also suggests different, and sometimes more elementary, proofs of important theorems and the possibility of generalization. We have given a proof of the classical Cayley–Hamilton theorem which illustrates that it is really a theorem about weighted digraphs; this was first noted by Rutherford [70] (see also [75], [81], and [7]). We have also seen how a large part of the proof of the Jordan canonical form—starting from Jacobi's theorem that a matrix is similar to a triangular matrix—can be made graph-theoretical (see [6] and the reference to Turnbull and Aitken there).

The theory of positive, more generally nonnegative, matrices—the so-called Perron–Frobenius theory—depends substantially on the zero-nonzero pattern of a matrix, and this translates to the digraph. For instance, an irreducible matrix becomes a strongly connected digraph; for more on this, one may consult [3] and [7]. We have seen how properties of the eigenvalues of a nonnegative matrix heavily depend on the digraph of the matrix. By use of the digraph, we were able to give a substantial generalization of the Geršgorin inclusion region for the eigenvaues of a matrix (see [7] and [78] for a lot more on this topic).

The permanent is an important and fundamental combinatorial function for which we have given a graphical interpretation. It can be used to motivate the important notion of an SNS-matrix. More about SNS-matrices and related topics can be found in [10].

We have included in the bibliography several books for further study and historical information as well as several classical papers that influenced the development of graph theory as a tool in matrix theory. The book [23] (in Serbian) also contains related and additional information.

Finally, we remark that the following four groups of papers, from electrical engineering, mathematics, and chemistry, belong to the origins of our combinatorial approach to matrix theory.

1. Electrical engineers have developed a series of methods for solving systems of linear algebraic equations, which appear in the theory of electrical circuits, control theory, and other areas. These methods use flow graphs (Coates [13], [26], [15]), signal flow graphs (Mason [59], [60], [80]), and Chan graphs (Chan and Mai [16],

[14]); the first two graphs are described in Chapter 6. It is remarkable that the mentioned graphs (and, especially, Mason's signal flow graphs) give a better insight into the physical system under description than the corresponding system of equations does. Therefore these graphs were introduced and used intuitively, the theoretical background of them often having been given later. The terminology used in this book is partially based on the terminology used in the literature of electrical engineering.

2. There are many mathematical papers in which results from the matrix theory are obtained or proved by graph-theoretical means. The founder of modern graph theory, Hungarian mathematician D. König, was the first who used graphical methods in matrix theory [55], [56], although even before König there were some attempts in this direction (see [64], footnote on p.260, where the Cauchy rule for the determination of the sign of a term in the development of a determinant is mentioned). See also more recent papers that belong to this group ([30], [20], [32], [33]), representing only a few examples. It is interesting to note that only papers with original and sufficiently nontrivial results obtained by graph theory were published, and it was only very recently that a few papers were published in which more elementary but more fundamental questions were also interpreted in this way ([22], [37], [38], [58]).

3. In the theory of graph spectra (see, for example, [18], [71], and [24]), including its applications to chemistry and to other branches of sciences, the results of matrix theory are used for investigations of graphs. Although we have here just an inverse procedure, compared with that of this book, a great number of results contributed to realize how, in the other direction, graphs can be used in matrix theory.

4. Some problems in electrical engineering, and in engineering in general, lead to the need of considering systems of linear equations whose matrix is sparse and entries are given numerically. Special methods of treating such matrices use graph-theoretical means to a great extent [11], [4], [76].

Answers and Hints

We give (partial) solutions or hints to a few selected exercises.

Chapter 1 Exercises

3. We have $kn = 2e$, where e is the number of edges.

9. In forming an even combination of $\{1, 2, \ldots, n\}$, one has two choices for each of $1, 2, \ldots, n-1$ (put in the combination of leave it out). When one gets to n, there is only 1 choice (put n in if an odd number of integers has already been taken; leave n out if an even number has been taken). This gives 2^{n-1} even combinations.

13. Use the fact that 99 is -1 modulo 100.

14. A basis consists of the $n-1$ vectors

$$(1, -1, 0, \ldots, 0), (1, 0, -1, 0, \ldots, 0), \ldots, (1, 0, \ldots, 0, -1).$$

Chapter 2 Exercises

4. Let $A = [a_{ij}]$ and $B = [b_{ij}]$ be upper triangular matrices of order n, so that all entries below the main diagonal equal 0. If there is an edge with nonzero weight in their König digraphs from black vertex i to white vertex j, then $i \leq j$. Now draw the composition $G(A) * G(B)$ to see that in $G(A) \cdot G(B) = G(AB)$ a similar property holds, implying that AB is also upper triangular. A similar argument works for lower triangular, or now use matrix transposiiton.

8. The König digraph of $I_n(i,j)$ has edges from black vertex i to white vertex j and from black vertex j to white vertex i. For $p \neq i,j$ there is also an edge from black vertex p to white vertex p. The identities now follow by examining the König digraphs $G(I_n(i,j)^2) = G(I_n(i,j)) \cdot G(I_n(i,j))$ and $G(I_n(i,k)I_n(k,j)I_n(j,i)) = G(I_n(i,k)) \cdot G(I_n(k,j)) \cdot G(I_n(j,i)$.

10. The product equals

$$\left[\begin{array}{c|c} O_2 & 2I_2 \\ \hline -I_2 & O_2 \end{array}\right].$$

Chapter 3 Exercises

2. The solution is easily obtained by using the digraphs $D(A)$ and $D(B)$ as drawn in Figure I:

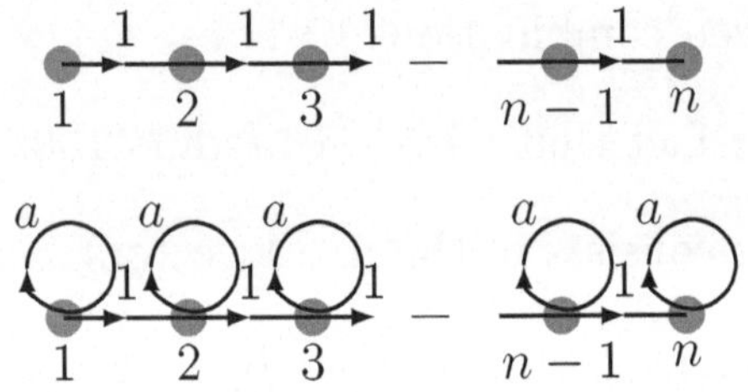

Figure I

8. Use the fact that for the permutation matrix P in the definition of a circulant of order n, $P^T = P^{n-1}$.

Chapter 4 Exercises

1. If $n = 2$, the matrix A has the form

$$\begin{bmatrix} 0 & a_{12} & 0 & a_{14} & 0 \\ a_{21} & 0 & a_{23} & 0 & a_{25} \\ 0 & a_{32} & 0 & a_{34} & 0 \\ a_{41} & 0 & a_{43} & 0 & a_{45} \\ 0 & a_{52} & 0 & a_{54} & 0 \end{bmatrix}.$$

The only edges (of nonzero weight) in the digraph $D^*(A)$ join odd numbered vertices to even numbered vertices (a directed bipartite graph). Hence no linear subdigraph exists, implying that $\det A = 0$.

Alternatively, one can consider the König digraph $G(A)$ and observe that the only edges go from even numbered black vertices to odd numbered white vertices, and from odd numbered black vertices to even numbered white vertices. Since there are $n + 1$ odd numbered black vertices and n even numbered white vertices, there is no 1-factor with nonzero weight; hence $\det A = 0$.

3. Let $G_n = G_n(a_1, a_2, \ldots, a_n)$ be the König digraph of the given matrix $\Delta_n(a_1, a_2, \ldots, a_n)$. Let $\mathcal{F}_n = \mathcal{F}_n(a_1, a_2, \ldots, a_n)$ be the collection of 1-factors of $G_n(a_1, a_2, \ldots, a_n)$, and let $\mathcal{F}_{n,k}$ be the collection of 1-factors with exactly k loops for $k = 0, 1, \ldots, n$. Then $\det \Delta_n(a_1, a_2, \ldots, a_n)$ equals

$$\begin{aligned} &\sum_{F \in \mathcal{F}_n} (-1)^{q(F)} w(F) \\ &\quad = \sum_{k=0}^{n} \sum_{F_k \in \mathcal{F}_{n,k}} (-1)^{q(F_k)} w(F_k) \\ &\quad = \sum_{k=0}^{n} \sum_{\{i_1, i_2, \ldots, i_k\} \in \{1,2,\ldots,n\}} a_{i_1} a_{i_2} \cdots a_{i_k} \sum_{F'_k \in \mathcal{F}'_k} (-1)^{q(F'_k)} w(F'_k), \end{aligned}$$

where $\mathcal{F}'_k$ represents the collection of 1-factors of $G_{n-k}(0, 0, \ldots, 0)$. This implies that $\det \Delta_n(a_1, a_2, \ldots, a_n)$

equals

$$\sum_{k=0}^{n} \det \Delta_{n-k}(0,0,\ldots,0)e_k(a_1,a_2,\ldots,a_n),$$

where $e_k(a_1,a_2,\ldots,a_n)$ is the kth elementary symmetric function of $a_1,a_2,\ldots,a_n$. It is easy to show that $\det \Delta_l(0,0,\ldots,0) = (-1)^{l-1}(l-1)$. Therefore $\det \Delta_n(a_1,a_2,\ldots,a_n)$ equals

$$\sum_{k=0}^{n}(-1)^{n-k-1}(n-k-1)e_k(a_1,a_2,\ldots,a_n).$$

4. Drawing the digraph $D^*(A)$ one sees that the only nonzero linear subdigraphs are those consisting of a cycle of length $i+1$ containing the first $i+1$ vertices and $n-i$ loops ($i = 0,1,\ldots,n$). Such a linear subdigraph contributes

$$(-1)^{1+(n-i+1)}(-1)^i a_i x^{n-i} = (-1)^n a_i x_{n-i}$$

to the Coates formula for the determinant. Hence

$$\det A = (-1)^n \sum_{i=0}^{n}(-1)^n a_i x^{n-i} = \sum_{i=0}^{n} a_x x^{n-i}.$$

6. The Coates digraph $D^*(A)$ cannot have any linear subdigraphs of nonzero weight; equivalently, the König digraph does not have any 1-factors of nonzero weight.

12. Observe that $A^T = -A$.

Chapter 5 Exercises

2. Drawing the digraph of a permutation matrix reveals its inverse.

9. Consider the digraphs $D^*(A), D^*(B), D^*(C)$ corresponding to the matrices A, B, C (see Figure II).

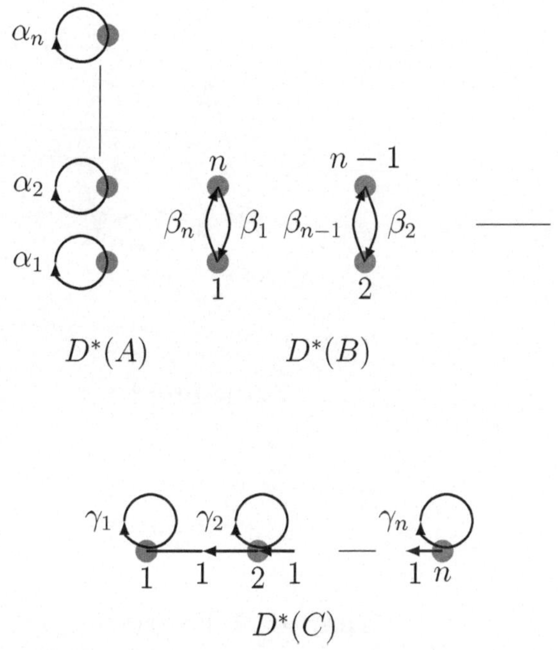

Figure II

Chapter 6 Exercises

8. The Coates digraph is drawn in Figure III. By inspection we get the following solution:

$$
\begin{aligned}
x_1 &= \frac{-A(eug+dvh)+B(avh-uhb)}{acvh+bfeu-bcuh-afve},\\
x_2 &= \frac{A(cvh+fve)}{acvh+bfeu-bcuh-afve},\\
x_3 &= \frac{A(feu-cuh)}{acvh+bfeu-bcuh-afve},\\
x_4 &= \frac{A(fvd+cug)+B(ubf-afv)}{acvh+bfeu-bcuh-afve}.
\end{aligned}
$$

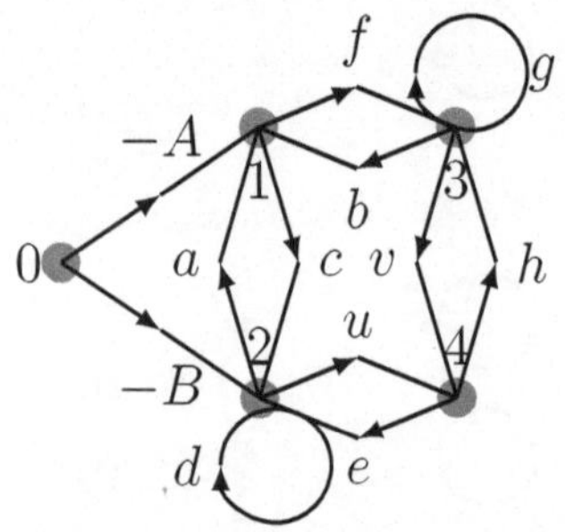

Figure III

Chapter 7 Exercises

5. Consider the cases $\lambda = 0$ and $\lambda \neq 0$ separately.

11. The characteristic polynomial is calculated using the Coates digraph of the matrix $\lambda I_n - A$ as drawn in Figure IV:

$$\begin{aligned} P_G(\lambda) &= (-1)^n \Big((-1)^n \lambda^{n-1}(\lambda - a_n) + (-1)^{n-1} a_1^2 \lambda^{n-2} + \\ &\quad + (-1)^{n-1} a_2^2 \lambda^{n-2} + \cdots + (-1)^{n-1} a_{n-1}^2 \lambda^{n-2}\Big) \\ &= \lambda^n - a_n \lambda^{n-1} - \lambda^{n-2} \sum_{i=1}^{n-1} a_i^2 = 0. \end{aligned}$$

The eigenvalues are

$$\frac{1}{2}\left(a_n \pm \sqrt{a_n^2 + 4\sum_{i=1}^{n-1} a_i^2}\right), 0\ (n-2 \text{ times}).$$

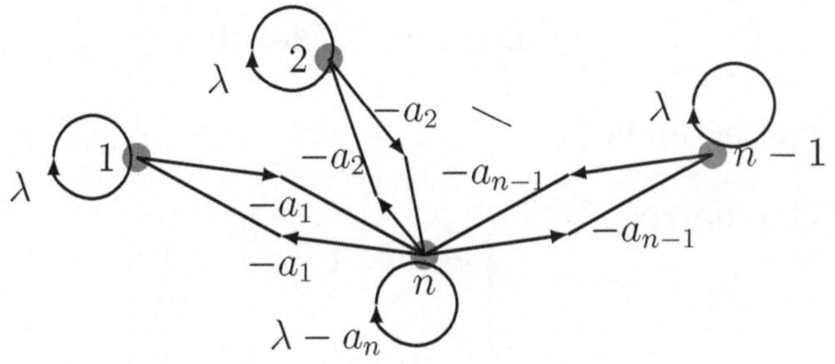

Figure IV

12. The number of different Jordan canonical forms equals the number of partitions of the integer 6:

$$6; 5, 1; 4, 2; 4, 1, 1; 3, 3; 3, 2, 1; 3, 1, 1; 2, 2, 2;$$

$$2, 2, 1, 1; 2, 1, 1, 1, 1; 1, 1, 1, 1, 1, 1$$

and so equals 12.

Chapter 8 Exercises

4. The exponents are 26 and 25, respectively.

5. An irreducible matrix of order $n \geq 2$ must contain at least one nonzero entry in each row (and column) and so contains at least n nonzero entries. If there are only n nonzero entries, then its digraph is a circuit and hence not primitive.

8. The key is that the digraph of A has at least one loop.

11. The adjacency matrix has r 1's in each row and column.

14. The graphs given are cospectral.

Chapter 9 Exercises

8. The permanent is 2^{n-1}.

11. The matrix

$$\begin{bmatrix} -1 & 1 & 0 & 1 \\ -1 & -1 & 1 & 0 \\ 0 & -1 & -1 & 1 \\ -1 & 0 & -1 & -1 \end{bmatrix}$$

is an SNS-matrix.

Chapter 10 Exercises

1. The adjacency matrix has two identical rows and so its determinant equals 0.

Bibliography

[1] A. C. Aitken, *Determinants and Matrices*, Edinburgh, United Kingdom: Oliver and Boyd, 1939.

[2] A. T. Balaban, ed., *Chemical Application of Graph Theory*, London: Academic Press, 1976.

[3] R. B. Bapat, T. E. S. Raghavan, *Nonnegative Matrices and Applications*, Cambridge: Cambridge University Press, 1997.

[4] V.A. Barker, editor, *Sparse matrix techniques*, Advanced course held at the Technical University of Denmark, Copenhagen, August 9–12, 1976, Lecture Notes Math. No. 572, Berlin-Heidelberg-New York: Springer-Verlag, 1977.

[5] S. Barnett, London-New York-Toronto, Ontario: *Matrices in control theory*, 1971.

[6] R. A. Brualdi, The Jordan canonical form: an old proof, *Amer. Math. Monthly*, 94(1987), No. 3, 257–267.

[7] R. A. Brualdi, H. J. Ryser, *Combinatorial Matrix Theory*, Cambridge: Cambridge University Press, 1991; reprinted 1992.

[8] R. A. Brualdi, *Introductory Combinatorics, fourth edition*, Upper Saddle River NJ: Pearson Prentice-Hall, 2004.

[9] R. A. Brualdi, *Combinatorial Matrix Classes*, Cambridge: Cambridge University Press, 2006.

[10] R. A. Brualdi, B. L. Shader, *Matrices of Sign-Solvable Linear Systems*, Cambridge: Cambridge University Press, 1995.

[11] J. R. Bunch and D. J. Rose, eds., *Sparse matrix computations*, New York-San Francisco-London: Academic Press, 1976.

[12] A. Cayley, A theorem on trees, *Quart. J. Math.* 23(1889), 376–378.

[13] C. L. Coates, Flow graph solutions of linear algebraic equations, *IRE Trans. Circuit Theory*, CT-6(1959), 170–187.

[14] S. P. Chan, *Introductory Topological Analysis of Electrical Networks*, New York: Holt, Rinehart and Winston, 1969.

[15] S. P. Chan, B. H. Bapna, A modification of the Coates gain formula for the analysis of linear systems, *Int. J. Control*, 5 (1967), No. 5, 483–495.

[16] S. P. Chan, H. N. Mai, A flow-graph method for the analysis of linear systems, *IEEE Trans. Circuit Theory*, CT-14 (1967), No. 3, 350–354.

[17] W. K. Chen, *Applied Graph Theory*, Amsterdam-London: North-Holland, 1971.

[18] L. Collatz, *Eigenwertaufgaben mit technischen Anwendungen*, Leipzig: Akademische Verlagsgesellschaft. Geest & Portig, 1963.

[19] L. Collatz, U. Sinogowitz, Spektren Endlicher Grafen, *Abh. Math. Sem. Univ. Hamburg*, 21(1957), 63–77.

[20] C. D. H. Cooper, On the maximum eigenvalue of a reducible non-negative real matrix, *Math. Z.* 131(1973), 213–217.

[21] C. A. Coulson, B. O'Leary, R. B. Mallion, *Hückel Theory for Organic Chemists*, London: Academic Press, 1978.

[22] D. M. Cvetković, The determinant concept defined by means of graph theory, *Mat. Vesnik* 12(27)(1975), 333–336. Dutch version: Definitie en berekaning van determinanten met behulp van grafen, *Nieuw Tijdsch. v. Wisk.* 63(1976), No. 4, 209–215.

[23] D. Cvetković, *Combinatorial Matrix Theory, with Applications to Electrical Engineering, Chemistry and Physics*, (in Serbian), Beograd: Naučna knjiga, 1980; 2nd ed. 1987.

[24] D. Cvetković, M. Doob, H. Sachs, *Spectra of Graphs—Theory and Application*, Deutscher Verlag der Wissenschaften–Academic Press, Berlin–New York 1980; 2nd ed. 1982; Russian translation: Naukova Dumka, Kiev, 1984; 3rd ed. Heidelberg-Leipzig: Johann Ambrosius Barth, 1995.

[25] D. Cvetković, I. Gutman, The algebraic multiplicity of the number zero in the spectrum of a bipartite graph, *Mat. Vesnik*, 9(24)(1972), 141–150.

[26] C. A. Desoer, The optimum formula for the gain of a flow graph or a simple derivation of Coates' formula, *Proc. IRE* 48(1960), 883–889.

[27] J. R. Dias, *Molecular Orbital Calculations Usig Chemical Graph Theory*, Berlin-Heidelberg-New York: Springer-Verlag, 1983.

[28] R. C. Dorf, *The Electrical Engineering Handbook*, Boca Raton, FL: CRC Press, 1999.

[29] I. S. Duff, A. M. Erisman, J.K. Reid, *Direct Methods for Sparse Matrices*, 2nd edition, New York: Oxford Science Publishers, The Clarendon Press, Oxford University Press, 1989.

[30] A. L. Dulmage, N. S. Mendelsohn, in Graphs and matrices, *Graph Theory and Theoretical Physics* (Editor: F. Harary), London-New York Academic Press, 1967, 167–227.

[31] S. Even, *Graph Algorithms*, London: Pitman Publishing Limited, 1979.

[32] M. Fiedler, Some applications of the theory of graphs in matrix theory and geometry, *Theory of Graphs and its Applications*, Proc. Symp. held in Smolenice in June 1963, Praque 1964, 37–41.

[33] M. Fiedler, Inversion of bigraphs and connections with the Gauss elimination, *Graphs, Hypergraphs and Block Systems*, Proc. Symp. on Comb. Analysis, held in Zielona Góra, September 1976, Zielona Góra, 1976, 57–68.

[34] G. F. Frobenius, Über zerlegbare Determinanten, *Sitzungsber. Preuss. Akad. Wiss.*, Berlin, 1917, 274–277.

[35] F. R. Gantmacher, *Theory of Matrices*, I, II, New York: Chelsea, 1960, (Translated from Russian).

[36] A. Graovac, I. Gutman, N. Trinajstić, *Topological Approach to the Chemistry of Conjugated Molecules*, Berlin-Heidelberg-New York: Springer-Verlag, 1977.

[37] J. V. Greenman, Graphs and determinants, *Math. Gazette*, 60(1976), 241–246.

[38] J. V. Greenman, Graphs and Markov Chains, *Math. Gazette* 60(1976), 49–54.

[39] J. L. Gross, J. Yellen, *Graph Theory and its Applications, 2nd ed.*, Boca Raton, FL: CRC Press, 2005.

[40] I. Gutman, Chemical graph theory—The mathematical connection. In eds. S. R. Sabin, E. J. Brändas *Advances in Quantum Chemistry 51*, Amsterdam: Elsevier, 2006, 125–138.

[41] I. Gutman, The acyclic polynomial of graph, *Publ. Inst. Math. (Beograd)*, 22(1977), 63–69.

[42] I. Gutman, O. E. Polansky, *Mathematical concepts in organic chemistry*, Berlin-Heidelberg-New York: Springer-Verlag, 1986.

[43] I. Gutman, N. Trinajstić, Graph theory and molecular orbitals, *Topics Curr. Chem.*, 42(1973), 49–93.

[44] F. Harary, A graph theoretic method for the complete reduction of a matrix with a view toward finding its eigenvalues, *J. Math. Phys.*, 38(1959), 104–111.

[45] F. Harary, The determinant of the adjacency matrix of a graph, *SIAM Rev.*, 4(1962), 202-210.

[46] L. Hogben, editor, *Handbook of Linear Algebra*, Boca Raton, FL: CRC Press, 2007.

[47] R. A. Horn and C. R. Johnson, Cambridge: *Matrix Analysis*, Cambridge University Press, 1985.

[48] R. A. Horn and C. R. Johnson, *Topics in Matrix Analysis*, Cambridge: Cambridge University Press, 1991.

[49] W. Jackson, A characterization of the reducibility of nonnegative matrices, *Proc. 8th South-Eastern Conf. on Combinatorics, Graph Theory and Computing*, Winnipeg, 1977, 381–384.

[50] D. Janežič, A. Miličević, S. Nikolić, N. Trinajstić, *Graph Theoretical Matrices in Chemistry*, Univ. Kragujevac, Kragujevac, 2007.

[51] M. Kac, Can one hear a shape of a drum?, *Amer. Math. Monthly*, 73(1966), April, Part II, 1–23.

[52] P. W. Kasteleyn, Graph theory and crystal physics. In *Graph Theory and Theoretical Physics*, ed. F. Harary, London-New York, 1967, 43–110.

[53] G. Kirchhoff, Über die Auflösung der Gleichungen, auf welche man bei der Untersuchung der Verteilung galvanischer Ströme geführt wird, *Ann. Phys. Chem.*, 72(1847), 497–508.

[54] W. Kocay, D.L.Kreher, *Graphs, Algorithms, and Optimization*, Boca Raton, FL: CRC Press, 2004

[55] D. König, Über Graphen und ihre Anwendungen auf Determinanten-theorie und Mengenlehre, *Math. Ann.*, 77(1916), 453–465.

[56] D. König, *Theorie der Endlichen und Unendlichen Graphen*, Leipzig: Akad. Verlag M.B.H., 1936. New York: Chelsea, 1950.

[57] P.-M. Lin, *Symbolic Network Analysis*, Amsterdam: Elsevier, 1991.

[58] E. K. Lloyd, Matrices, graphs and adjoints, *Math. Gazette*, 61(1977), 201–204.

[59] S. J. Mason, Feedback theory, some properties of signal flow graphs, *Proc. IRE*, 41(1953), 9, 1144–1156.

[60] S. J. Mason, Feedback theory, further properties of signal flow graphs, *Proc. IRE*, 44(1956), 7, 920–926.

[61] M. Milić, Flow-graph evaluation of the characteristic polynomial of a matrix, *IEEE Trans. Circuit Theory*, CT-11(1964), 423–424.

[62] H. Minc, *Nonnegative Matrices*, New York: Wiley, 1988.

[63] E. W. Montroll, Lattice statistics, *Applied Combinatorial Mathematics*, ed. E. F. Beckenbach, New York–London–Sidney: John Wiley and Sons, 1964, 96–143.

[64] T. Muir, *The Theory of Determinants in the Historical Order of Development, I*, New York: Dover, 1960.

[65] M. Parodi, *La localisation des valeurs caractéristiques des matrices et ces applications*, Paris: Gauthier-Villars, 1959.

[66] J. K. Percus, *Combinational Methods*, Berlin-Heidelberg-New York: Springer-Verlag, 1969.

[67] M. Perić, I. Gutman, J. Radić-Perić, The Hückel total π-electron energy puzzle, *J. Serb. Chem. Soc.*, 71 (2006) 771–783.

[68] K.H. Rosen, editor, *Handbook of Discrete and Combinatorial Mathematics*, Boca Raton, FL: CRC Press, 2000.

[69] D. E. Rutherford, Some continuant determinants arising in physics and chemistry, *Proc. Roy. Soc. Edinburgh Sec. A*, 62 (1947), 229–236.

[70] D. E. Rutherford, The Cayley-Hamilton theorem for semirings, *Proc. Roy. Soc. Edinburgh Sec. A*, 66 (1964), 211–215.

[71] H. Sachs, Beziehungen zwischen den in einem Graphen enthaltenen Kreisen und seinem charakteristishen Polynom, *Publ. Math. (Debrecen)* 11(1963), 119–134.

[72] H. Schneider, The concept of irreducibility and full indecomposability of a matrix in the works of Frobenius, König and Markov, *Linear Algebra and Appl.*, 18 (1977), 139–162.

[73] C. E. Shannon, The theory and design of linear differential equations machines, OSDR Rep. 411, U.S. National Defense Committee, 1942.

[74] V. Strassen, Gaussian elimination is not optimal, *Numer. Math.*, 13(1969), 4, 354–356.

[75] H. Straubing, A combinatorial proof of the Cayley–Hamilton theorem, *Discrete Math.*, 43(1983), 273–279.

[76] R. P. Tewarson, *Sparse matrices*, New York-London: Academic Press, 1973.

[77] N. Trinajstić, *Chemical Graph Theory*, Boca Raton, FL: CRC Press, 1983; 2nd revised ed., 1993.

[78] R. S. Varga, *Geršgorin and His Circles*, Berlin-Heidelberg-New York: Springer-Verlag, 2004.

[79] D. B. West, *Introduction to Graph Theory, 2nd edition*, Upper Saddle River, NJ: Prentice-Hall, 2001.

[80] D. Younger, A simple derivation of Mason's gain formula, *Proc. IEEE*, 51(1963), 1043–1044.

[81] D. Zeilberger, A combinatorial approach to matrix algebra, *Discrete Math.*, 56(1985), 61–72.

[82] F. Zhang, *Linear Algebra: Challenging Problems for Students*, Baltimore: Johns Hopkins University Press, 1996.

[83] R. Zurmühl, *Matrizen und ihre technischen Anwendungen*, Berlin-Göttingen-Heidelberg: Springer-Verlag, 1961.

Index